The Secret of the Universe

by Nathan R. Wood, [1932]

The Secret of the Universe

"God, Man and Matter"

By NATHAN R. WOOD

New York: F. H. Revel & Co.

[1932]

Edit By B.A. Bey

INTRODUCTION

WHEN the Hebrew Singer, marveling at the fact of which he had no doubt, that man is visited by God, exclaimed "When I consider Thy Heavens," he was undoubtedly comparing the apparent littleness of man with the vastness of the universe of which he was conscious. The expression "Thy heavens" included for him that universe whose existence was demonstrated by the "Moon and the stars." Evidently it was a song composed at night. Like all inspired expressions it was inclusive. How much it included, the Singer certainly did not know. How vast the universe is, we even now do not know. But to-day we know its vastness far better than David did. Science bewilders us with its affirmations as to its extent. This enlargement of apprehension does but accentuate for thinking minds the wonder of it, and creates what we now often describe, borrowing Haeckel's words, as "The Riddle of the Universe." To recognize a riddle, is to postulate a solution. Finally, somewhere, somehow, there must be a whole, which resolves all parts into itself. Perhaps we should say--that is the conviction of faith, which is the last word of reason. The mind of man has been seeking ever for a statement of that Whole in terms which include all the parts, in other words asking "What is the Secret of the Universe?" For every suggestion which Philosophy has made we are grateful, for its blunders which have needed correction and so have helped us, as well as for its recognitions of principles and laws which abide.

To that great Quest President Nathan Wood has made a contribution in this book. To me it is more than a Quest, it is a Conquest. In bulk this book is small. In reach it is as vast as the Universe. I could describe it as "The Philosophy of the Universe." The Author probably would prefer to call it "A Philosophy of the Universe." Whether "The," with its suggestion of finality, or "A," calling for further investigation, it is Philosophy, and must make an appeal to all who are seeking light on the fascinating mystery.

Personally when Dr. Wood did me the honor of allowing me to peruse the manuscript, I read it with growing amazement, at the final inclusiveness of its outlook, and the clear and cogent statement of its examination of those parts which interpret the whole. The first three parts of the book are as fascinating as a Fairy Story. That perhaps is an unfortunate figure of speech as it may suggest something fanciful. I employ it rather in the sense that truth is stranger than fiction. The final part requires more careful, or shall I say slower reading, as the Author gives us a keen and penetrative analysis of the Universe.

It is not the work of an introduction to give away a "secret." That must be discovered by the reader of the book. At least I do not hesitate to say here is a book, startling, challenging, scholarly, sane, courteous; and it must make its appeal to the consideration of those who are not content to drift through life taking things as they are, but desire to challenge life, not only in its passing hours, and nearest dust, but in its vastness and entirety. In this age, when Philosophy has really had nothing new to say for many generations, but has been satisfied with garbing in new terminology the thinking of other days, here is a book which will surely make men stop-look-and-listen. I would like to put a copy first in the hands of every theological student, and then also in the hands of every student who is endeavoring to "beat his music out."

G. CAMPBELL MORGAN.

CONTENTS

INTRODUCTION. THE SECRET OF ALL THINGS

CONCLUSION. THE SECRET OF THE UNIVERSE AND THE RIDDLES OF THE UNIVERSE

PART I. THE PATTERN OF THE UNIVERSE

I. THE OUTER UNIVERSE

Why is the universe what it is?--Is it really a universe?--What is its structure?--The basic things--Space--Matter--Time--"Is there a universal principle?"--The Equation of the Universe--How far does the structure of the universe extend?--How the being of God was presented to us--The Data classified--What Space is--What Matter is--What Time is--The Structure of the Universe--How the cause explains the structure--How the structure confirms the cause--The vision of the universe.

WHY is the universe what it is? That is the riddle of the universe. Can we ever solve it? Can we grasp it at all? Of course we cannot with human minds reach out to the ever-receding infinities of universe beyond universe of stars. Neither can we reach inward to the equal infinities of world within world in the atom. This does not need words. We know that we can never do it. We cannot grasp what it really means that a certain island universe is millions of light-years away. We cannot grasp what it means that the electron moves in its orbit around the proton in the atom a quadrillion times a second. It is no shame to us that we cannot grasp such things as these. Our minds are not geared to the infinite. If they were, could we harness them any more to the ledger, the plough, the tool-chest or the cook-stove? What would it profit to grasp the nebula and the electron, and starve or freeze? But our minds do seem fitted to understand. They can apparently understand the quality and meaning of things whose immensity they cannot grasp. They are evidently fitted to understand everything which can be understood. That seems somehow to be what they are for.

Then can we understand the universe? Can we find why it is what it is? That surely comes first in understanding it. Can we find an organic reason in it, and see it as a whole? Some put it in this way:--"Is there a formula of the universe? And can the formula be known?" It is worth every effort to understand this force or that fact in the world around us. What then is it worth to understand the universe!

Why is the physical world just what it is, and not something quite different? Could the physical world have existed in some other form and order? If it could have been wholly different, why is it what it is? Or if it had to be what it is, why did it have to be so? Is there a reason? What is the reason?

It is not enough to say that all things evolved into this present form of things out of a simpler condition. For if they did evolve, why did they evolve into this order of things, and not into something quite different? That is our question. Or if they came at a creative stroke into what they now are, why was it into just this form and character and not into some other?

Is this really a universe, then? That is, has it unity? Has it a structure, with a reason for the structure? For we surely do want, if we can, to see the universe as a whole. If we cannot see it all, which is a more than doubtful possibility, we can at least see what it all means. Can we see it as a universe, a genuine unity, including the world of matter and the world of mind? Is it such a universe, with an organic pattern and principle in it all? In other words, is there a structure of the universe, with a reason for it? What is that structure?

If we had not asked these things before, modern science would drive us to ask them now. This physical universe, this vast fabric of forces, this interplay of laws and energies, why is it so precisely and accurately what it is? There must be a structure of the modern universe. And if there is such a structure, there must be a reason for it, a universal reason why it is just what it is. What is that reason?

To find the structure of the physical universe, we do not need more complicated knowledge. We need to simplify. We need to use what we have. We must find the basic things, which form the universal structure.

What are the basic things? Those things which lie back of all other things? Those things which are the basis of all other things, and include all other things, and exist in all other things? Those will be the basic things. They should not be very hard to find. When we have found them we shall be on our way toward understanding the riddle of the universe. We must find the basic things.

Space

The first one may be agreed upon without difficulty. It is Space--the basic thing in the physical universe. It is back of every other basis of the physical universe. We may have many different views about space, and many speculations. We may think of space as an outward reality, or as our way of seeing the universe. But, in any case, space comes first. Upon one thing we can all agree. Whatever each of us may mean by space, this is, as we all know, a space universe.

What is space? Of what does it consist?

We may speculate much about this. But again we can without difficulty agree on the essential thing. Space as we all know it and live in it and use it consists of three things. We call these three things three dimensions, or three directions. We name them generally length, breadth and height.

Two words may be objected to nowadays, when we talk about "three dimensions." One word is "dimensions," and the other is "three." But for our simpler and basic purpose neither objection will be made.

It may be suggested, and rightly, that the term "dimensions," the term most commonly used, and which therefore we are using, means measurements, and therefore implies limited distances. "Length, breadth and height," too, may be taken as limited terms. "Length" may mean the distance between two definite points, which means a limited distance. "Breadth" also may be used as a term of measurement, rather than a term of unlimited direction. "Height" also may signify a limited distance upward. These terms are not wholly unambiguous when we would signify unlimited space. It would be more unmistakably accurate to say "directions" instead of "dimensions." Space has three "independent directions," the mathematician says. Those words do not carry the possible meaning of measurement, as "dimensions" do. But it is difficult to name the three directions. Shall we call them "north," "east" and "up?" Or x, y and z? These are accurate. But people in general are not accustomed to these terms. They are on the contrary accustomed to saying "dimensions" when they mean unlimited "directions," and to saying "length," "breadth" and "height" when they mean the three general "directions" of unlimited space. May we not, then, use the terms to which most readers are accustomed, so accustomed that these terms are second nature to them when they think about the directions of space. But may we mean by these terms, not the limited measurements of a box, or house, or geometric figure, but the unlimited "dimensions" of free space. That is what the mathematician means by "three independent directions," and it is what the ordinary person means by the "three dimensions of space," and it is what the mathematician means when he talks about a "fourth dimension."

In this sense of the terms, space as we know it and live in it and use it consists of three things. They are three dimensions,--length, breadth and height. As we have said, we may speculate about space. Mathematicians may demonstrate a fourth dimension and fifth dimension, and some profound realities may or may not reside in the demonstrations. But when we build a house we build it in three dimensions. No man in the world has ever raised a cabin or a cathedral of either more or less than three dimensions. No thinker would know how to plan a structure of more than three dimensions. Whatever the refinements, the subtleties, of space may be, it is clear that the basic space, the space of common knowledge and experience, is of three dimensions. It consists of those dimensions. It is length, breadth and height. This is the first thing in the structure of the physical universe.

Matter

What else does the universe around us tell us of the basic things? What else is the structure of the physical universe? Of what else does it consist?

That too is a matter of common knowledge and experience. The man who moves about unthinking in the physical world and the man who spends restless days and sleepless nights in exploring the secrets of the physical world are at one in this.

Next to space is that which fills space and embodies space, and gives to us all the phenomena of the physical universe. We call it matter. But we know now that it is primarily energy. We can agree to call it matter, if by matter we mean that form which energy takes so that we can see it, or hear it, or feel it. We mean all of that which fills and gives outward reality to space. We all of us take this for granted in all of our daily thinking and activity.

What is the nature then of this which occupies space, and makes a visible, audible, tangible universe? Of what does it consist?

Here again we are among things which we know. For even if we avoid hypothesis and speculation there is much which now we definitely know.

Modern physics and chemistry find, first and basic in matter:--energy,--vast, unknown, unseen, a primal thing, out of which all things in the physical universe come.

We may, it is true, define energy as "mass multiplied by the square of the velocity." That is the technical velocity. Rather the energy is itself the cause. Mass may be, and doubtless is, simply a manifestation of that energy. And the velocity is surely a result, a manifestation, of that energy. "Mass times the square of the velocity" then is not what energy is. It is the way we measure energy. We measure energy by its manifestations, of mass and velocity.

Second, modern physics and chemistry find, growing out of energy, embodying energy,--motion,--that great, unceasing, unresting motion, which fills and which is the physical universe.

Third, they find all those infinite complexities and variations of motion, those varying velocities, into which motion differentiates itself, and which, when they present themselves to us as waves of light, of air, of sound, we recognize as physical phenomena, light, color, sound, heat, cold, hardness, softness, scent, moisture, dryness. They are not dependent upon our recognition or experience of them. They register themselves upon mechanical instruments as readily as upon human senses, showing that they exist apart from human beings and human perceptions. They are probably not different "kinds" of motion. They are probably, as we know that light waves and sound waves are, simply different rates of motion, or different velocities into which motion differentiates itself. We call them phenomena. We think of them in connection with our senses, because that is the way in which we become acquainted with them. But they definitely exist apart from our senses. If we remember that they are in themselves differentiations of motion, which exist apart from us, we may call them, as we know them, phenomena.

This is that universe of matter or substance in which we are, and of which we are a part. It consists in the most literal sense of these three things,--energy, motion and phenomena.

Time

Is anything else basic in the physical universe? Is there anything else which is of the very structure of the universe? Yes. There is one other thing of which we can be sure. There is one other thing which every thinker agrees to recognize as an absolute basis of the physical world. That is,--Time. Whatever our view of Time may be, we regard Time as basic. In eternity everything must doubtless be timeless. But in the world nothing is timeless. Time is of the essence of everything in the physical universe. This is, by universal agreement, a time universe.

When we ask what time is, and how we may resolve it into its component parts, the answer is simple. We need not speculate how far time is an outer reality, and how far it is our way of conceiving things. For whichever it is, or if, as is doubtless the case, time is both an outer reality and our way of conceiving things, the facts about time are so universal, so clear-cut and so familiar as to leave no practical question at all in any mind. Time, as a matter not of speculation but of simple experience, consists of three great constituents,--past, present and future. We all know them. We all live in them. They include everything, and make this a time universe.

Is There a Universal Principle?

Is there any other basic element in the physical universe? Is there any other thing, which is not to be resolved into one or all of these three,--space, matter and time? No. We know of no other. These are what all can agree upon.

This is a universe. Of that we feel very sure. Is there then any universal principle? Is there anything which space and matter and time have in common, beside the fact that they are in the same universe? We need not speak now of their relationship to each other. That is a profound and subtle question. We are asking rather the simpler question, Is there anything which these three,--space, matter and time,--have in common? Is there anything like a universal structure in all three?

To this there is an immediate answer. There is one thing which these basic realities have in common. It suggests itself at once. It is, that each one of these elemental things of the physical universe is threefold.

It is length, breadth and height, in one Space.

It is energy, motion and phenomena, in one Substance.

It is past, present and future, in one Time. That is truly a vast coincidence.

As a space universe, as a substance universe, and as a time universe, it is in each case three things in one. This is at once the most obvious and the most striking thing about this structure of space, of matter and of time. Different as these three elements are, they have this in common. Each is three things in one. The riddle of the universe now takes this form:--Why is the physical universe, in each of its basic elements, three things in one?

Where is the answer? How far in the universe does this vast coincidence extend? Does it include man, who is so much a vital part of the universe? Does it include God, who is the ground of the universe? Does it reach as deep and as high as that? Is the answer to the riddle of the threefold universe to be found in God?

We make no apology for speaking of God. A universe without God is meaningless now. For the day of the blind soul in a black universe has gone by. The stars and the atoms have taught us to see. The deepest instinct of one who lives in the modern universe reasons irresistibly. "The fool," it may be, has always "said in his heart 'There is no God.'" But the modern universe leaves scant footing for such a dance of the morons. Even when the mind does not analyze, the heart reasons, and that is the deepest reasoning.

We gain, too, when we do not overstate that reasoning. Little is gotten by claiming infinite conclusions from finite premises. It is better to put things within bounds, and to reach certainties. Shall we try to do that?

The Equation of the Universe

From time to time in these days the world waits for new equations of the universe to express the newest discoveries. In the nature of things such an equation is hard to understand and sure to pass when newer things are found. But it is possible, one may believe, to work out an equation which shall be simple, self-evident and ageless.

This universe about us is vast beyond our comprehension. New universes of stars beyond this universe are floating into our ken. Apparently it is infinite. Certainly it is inconceivably vast.

The cause of it must be at least as great.

This universe, from the island universes quintillions of miles away to the electrons whirling in the invisible atom, is one immeasurably articulated, rationally working fabric.

The cause of it all must be at least as rational as that.

This universe contains personal beings, who think, who love, who hate, who hope, who fear, who choose, who determine.

The cause of such beings, of a universe which contains such beings, must be at least as personal as they.

The equation of the universe is clear. A vast, rational, personal cause of the universe = God.

How far then does the vast coincidence found in the threefold structure of space, the threefold fabric of matter, and the threefold existence of time, extend in the universe? Does it extend to God, who is the ground of the universe? Is the answer to the riddle to be found in Him?

There is an immediate and striking answer again. That is, that the greatest religion in the world presents God also as being threefold. It is the religion which coincides with modern civilization. It holds so great a place in the modern world that whether one agrees with it or not one cannot disregard it. That religion brings this striking answer. For the Christian religion, the Christian Bible and Christian consciousness present God as three in one.

Is it possible then that we have after all a vast structure of things which includes not only space and matter and time, but also, as it ought to do, includes God, who is the ground of space, of matter and of time? Is it the same three foldness?

That is not a conclusion to be jumped at. This is no place for easy inferences. We need some very careful thinking.

Are the threefoldness in space, in matter and in time, which are the structure of the physical universe, and the threefoldness presented by the Bible and Christian consciousness as the being of God, the same kind of threefoldness? Are these two,--what we may call the scientific threefoldness of the physical universe and the Biblical threefoldness of God,--so much alike that they are obviously the same thing, in different terms? If they are, we must face the question of a vast structure of things which includes God. If they are not the same, we need not consider the coincidence of such threefoldness in the physical world and in that presentation of God.

And when we say "Are they the same?" we should mean "Are they exactly the same?" It is a case for scientific precision, not for metaphors and similes.

It is a dramatic question. The threefoldness so presented as the being of God is mysterious and purely a thing of the spirit,--not having, as presented there, any effort whatever at likeness to the physical world. How can it be in any real sense the same thing as the structure of a physical universe?

Yet that is what scientific exactness would mean,--a likeness so close and so complete that the two would be, in different realms, manifestly the same principle.

We ask the question again, then, in view of that presentation of God. Does the vast coincidence found

in the threefold structure of space, the threefold fabric of matter, and the threefold existence of time, extend to God, who is the ground of the universe of space, and matter, and time?

We have sketched rapidly the threefold structure of space, of matter and of time. What now is that other threefoldness like, which is presented by the Bible and by the consciousness of Christians as the being of God?

Inductively, not dogmatically, we should find and phrase that threefoldness. We should go to the sources.

It is Jesus of Nazareth who presents the threefoldness of God in the Bible. What did he say? How did he bring the presentation?

The religion of the Old Testament had taught that God is one. That was a new message, a startling emergence, amid the sea of surrounding polytheism. Then, after many centuries, Jesus of Nazareth came, and began the presentation of God as three in one.

At the baptism of Jesus, at the beginning of his public life, the Bible declares, as the Son was praying, the voice of the Father spoke out of the open heavens: "This is my beloved Son," and the Spirit descended visibly upon him. In all his teaching and preaching, and in private conversation, Jesus constantly spoke of his Father and himself as two distinct persons, and yet declared equally "I and my Father are one," even naming himself first! At another time he said, "He who has seen me has seen the Father." On one occasion he is quoted as declaring "All things have been delivered unto me by my Father; and no one knows the Son, except the Father; neither does anyone know the Father, except the Son, and he to whom the Son wills to reveal him." In his last and longest recorded talk with his disciples, in the upper room, the evening before his crucifixion, he said, "The Holy Spirit, whom the Father will send in my name, he shall teach you all things, and bring to your remembrance all that I said unto you." Not many minutes later, in the same conversation, as he and the disciples walked out from the city to Gethsemane, before his betrayal, it tells us that he said, "When the Comforter is come, whom I will send unto you from the Father, even the Spirit of truth, who proceeds from the Father, he shall bear witness of me." In his last command he directed his disciples, wherever they should win men, to "baptize them into the name of the Father, and of the Son, and of the Holy Spirit," a name, but a threefold name. In these sayings he of course claimed to be God. That was why they crucified him. The fact that throughout his recorded ministry he made that claim, and that this assumption appears in everything that he said, is the basis of all the rest of the Biblical presentation of the threefoldness of God.

A third also is everywhere in the Bible represented as being God, and in the New Testament churches was treated, listened to and obeyed as a Divine Person. He is named there "the Holy Spirit," or simply in many cases "the Spirit." He is also called "the Spirit of Jesus." The Spirit was "sent," Jesus said, "by the Father," and was sent "in the name of the Son," and "by the Son, from the Father." Indeed everywhere in the New Testament each one of these three is represented as being God.

The thinking of many centuries has formulated the data found in the words of Jesus and the New Testament. It presents in logical form a thing so extraordinary and so mysterious that one may naturally question whether the threefoldness in the physical world can really have anything in common with it.

The Data Analyzed

1. Absolute Threeness. It is apparently an absolute threeness. Everything in the Biblical description of the three called Father, Son and Holy Spirit presents definitely and absolutely no more and no less than these three persons in the Godhead. It seems to be an absolute threeness. In such an absolute threeness each one of the three is so genuinely distinct from the others that no one of the three can possibly be either of the others. Otherwise they are not absolutely three. And in such threeness, in which there absolutely must be three, no two of the three therefore can exist without the third.

2. Absolute Oneness. The Three in that Trinity are represented as absolutely one. They are not only called so in direct statements, but further than that is the fact that each one is represented as God. That means that each one is not a part of God,--each one is God. It means that each one is the Whole. For God is not divisible. Personality is not divisible. If God is Three in One, each one of the three is God, and each one is the whole of God.

3. It is very clear that, as personal beings, Father, Son and Holy Spirit are represented as three things which God is. In other words, they are pictured as three modes of being. They do not mean primarily three ways in which He acts, three modes of action or manifestation, although of course if they are three things or persons which He is they present three ways in which He acts. But primarily, if God is Father, Son and Holy Spirit, it means three things which He is, or three modes of being.

4. The Scripture makes it invariably clear that in some way it presents the Father as first, the Son second, the Spirit third. It evidently does not mean that one is first in deity, when all are represented as being God. Neither does it mean that one is greater, when all must be infinite. Nor does it mean that one is first in time, when all are represented as eternal. It can only mean that one is first, one second, one third, in a logical, causal order.

5. In the Scripture the Father is represented as the Source. The eternal Son is "begotten, of the Father." The eternal Spirit "proceeds," from the Father, through the Son. Jesus said, "the Spirit whom I will send unto you, from the Father," and "the Spirit whom the Father will send in my name." This relation is mysterious, but emphatic.

6. In that extraordinary Trinity of Scripture the Father is unseen. He reveals Himself in the Son. "No man has seen God. The only-begotten Son, who is in the bosom of the Father, he has

revealed him." The Son is the visible embodiment of the Father, and of the Godhead. "He, being the express image of the Father,"--"he that hath seen me hath seen the Father,"--"In him," the Son, "dwelleth all the fulness of the Godhead bodily." The Son acts. He does the things which are done. He creates. It is he who becomes Man. It is he who dies, and who rises. It is he who raises the dead, and who judges. The Son, the Scripture says, works now among men through the Spirit. The Spirit, like the Father, is unseen. He reveals the Son. That is His chief work. And He reveals the Father, in the Son. He works unseen, in other beings, as for instance in man. This is the presentation by the Bible.

7. One thing there is, with which we should not attempt comparison in the physical world. The Biblical presentation of Father, Son and Holy Spirit in God means what can best be described as "three Personal centers of consciousness in one Being." That cannot possibly be paralleled in the impersonal physical world, and should not be sought there. But the other data which we have given include what the mind of man has fairly deduced through the ages from that presentation of threefoldness by Jesus and the New Testament.

How does the structure of the universe compare with this? It seems impossible that such abstruse, complex, mysterious, purely spiritual things could have a counterpart in the physical world. But such a comparison, and a very exact comparison, is the only way to answer the question whether that threefoldness of the physical universe coincides with the mysterious Trinity presented by the Bible as the being of God.

How Does Space Compare?

First, then, Space.

Space is height, length and breadth,--three things in one space.

Is it an absolute, necessary threeness? Yes. That answers itself. For, as we have said, in the basic space of common experience, in which all men move and act, there are three dimensions. There are no less, and no more. No one who builds, or runs, or flies, or who studies the stars, ever considers either more or less than three dimensions. Whatever there is of reality in the proofs of other dimensions, the elementary space is exactly and always of three dimensions.

Is it an absolute threeness by those deeper specifications, which would belong to such spiritual threefoldness as that described in the New Testament,--the specifications that no one of the three can possibly be either of the others, and that on the other hand no one of the three can exist without the others?

In the first place, are the three dimensions so absolutely distinct that no one of the three can possibly be either of the others? Yes, that specification is very evidently met. No one of the three can possibly be either of the others. Height cannot be length or breadth. Length cannot be confused with height or breadth. Breadth is not and cannot be either of the others. Thinking presents no clearer distinctions than these three. Indeed, the absolute, self-evident distinctness of the three elements in space is made a basis of the exact science of mathematics, and especially of that most exact science of geometry.

So necessary and so absolute is this threefoldness of space,--can we then apply to it the other test, a test almost spiritual? Are the three dimensions so necessary, is space so absolute a threeness, that without any one of the three the other two could not exist? That is readily tried out. That is the way to answer that query. Take away height. Then length and breadth become a plane surface. A plane surface is purely imaginary. It has no actual existence. Length and breadth as a plane surface, without height, do not really exist in the world of actual things. Mathematics may imagine them. But actual space does not exist unless it has all three directions or dimensions. With any one of the dimensions missing, space becomes imaginary and non-existent, and therefore its other two dimensions become imaginary and non-existent. To give existence to any one of them all three are necessary. Yes, even by these specifications, which seem more spiritual than physical, space is indeed an absolute threeness!

Is space an equally genuine oneness? Can it so resemble the absolute oneness which would be in Three Persons in One, the kind of oneness in which each must be the whole? That is a spiritual kind of unity which hardly seems to belong to the physical world. Does space show any such thing as that?

We can test it in the laboratory of the mind. Think carefully for a moment of these three dimensions of which space is composed. Are they so much one that each dimension is really the whole of space? Picture to yourself that dimension or direction of space which we call length, or forward and back. There are in space an infinite number of parallel lines all running forward and back. Now manifestly all possible points in space, all the myriad points which make up space, are contained in those parallel forward and back lines. Those lines, that dimension or direction, include all of space. There is nothing in space, no point in all its breadth and height, which is not included in its length. In a very definite sense its length comprises the whole of it. The length, it is true, does not exist without the breadth and height of space. For space does not exist unless it has all three elements. But its length clearly includes all its breadth and all its height. In a very remarkable and real sense the length is the whole. But the same may be said of the dimension or direction of breadth or side-to-side. Every point in all the length and height of space is contained in the infinite number of parallel lines of space which run from side to side. There is nothing in space, no point in all its length and height, which is not included in its breadth, in its side-to-side direction. In a very real and remarkable sense its breadth comprises the whole of space. But the same is true of the dimension or direction of height-and-depth, or up-and-down. All of the points which make up the length and breadth of space are included in the infinite number of parallel lines running up and down in space. There is nothing of space and its length and breadth which is not included in its dimension of height-and-depth, its direction of up-and-down. In a remarkable sense the dimension of height-and-depth is the whole of space. It is true that no one of the three,--length, breadth and height,--can exist apart from the other two elements. For space itself does not exist unless it has all three. But in a genuine and remarkable sense each of the three is the whole of space. The three are one in that kind of absolute oneness in which each is the whole. It is the extraordinary kind of oneness which is found in the Trinity of Scripture, where each one of the Three is not a part of God, but rather each one of the Three is God. So space, in which each direction or dimension is not a part of space but is the whole of space, is an absolute unity. It is as absolutely one as it is absolutely three. It is absolute triunity.

We may say of this triune space, so wholly three and so completely one, that it is so much a triunity that it is nothing else at all. It is bare triunity, reduced to its simplest terms, with all the characteristics of absolute triunity, and with no other characteristics. It has no evident characteristics but its threeness and its oneness. It is as though the Creator has chosen to make one thing which is sheer, simple, essential triunity! And on it and in it and of it He built His universe.

One other question should be asked. In that Triunity of Scripture the Three are not three things which God does. They are presented as three things which God is. They are modes of being. Is that true of space? Consider that too. Are length, breadth and height three things which space does?

No. Length, breadth and height are of course three things which space is.

Then, like the Threeness in the New Testament Triunity, the three in this absolute triunity which we call space are three modes of being.

This is, beyond question, a very remarkable coincidence. As far as space has any characteristics, it is pure triunity exactly like that Trinity described in the New Testament. In threeness, in oneness, as modes of being, the likeness is remarkable.

How Does Matter Compare?

But space is not a very physical thing. The second great element in the physical universe, matter, is much more physical. How does matter compare with such Triunity as the Bible describes? We know the threefoldness of matter well. There is energy, the primal thing. All sciences recognize it. Then there is motion, coming out of energy,--motion of waves, of electrons,--all the universe of motion everywhere. Then there are phenomena, through which motion in its varying velocities touches the senses,--light, sound, heat, hardness,--all the multitudinous impacts of motion upon sight, hearing, touch. They are in themselves varying speeds of motion, existing wholly apart from human senses. But they touch us as light, sound, heat, hardness, texture. Does, then, that threefoldness of energy, motion and phenomena show any likeness to the mysterious, abstruse and purely spiritual Triunity presented in the New Testament?

Only a very careful comparison will answer the question.

Is the threeness of energy, motion and phenomena such an absolute and inevitable threeness? That ought to be possible to discover.

Is it true that there can be no more than these three primary distinctions in matter? Certainly it seems to be so. These three cover the entire range of matter, from the energy in which it begins, to the final phenomena which strike upon our senses. They include everything which matter is. There seems truly to be no place for any other primary distinction, not included in these three.

Then is it true that there can be no less than these three? Are the three so distinct that no one of the three can be either of the other two, thus leaving two instead of three? That must be found by comparing the three. Is energy, for instance, the same as motion? Manifestly it is not. Energy is the source, the potentiality, the cause, of motion. It can pass into motion, and does do so. But it is not itself motion. Neither on the other hand is motion the same as energy. It contains energy, and embodies energy. But it is not the same as energy. Is energy the same as phenomena? That answers itself, too. Energy, always and inveterately unseen, is not the same as phenomena, by their very nature visible, audible or tangible. It issues, through motion, in phenomena, but it is not itself phenomena. Is motion the same as phenomena? It shows itself in phenomena. Its identification with them is close. But when we say "motion" we do not mean the same thing as when we say "phenomena." The idea is different because the thing is different. Nor, if energy and motion are neither of them the same as phenomena, are phenomena the same as energy or motion. No one of these three can be put aside, as being so much the same as one of the others that we do not need to give it a separate name.

Manifestly the energy-motion-phenomena substance of the universe, which we call matter, can be neither more nor less than three.

Then is matter such a threeness that no two of the three can exist apart from the third? Is that absolute threeness possible? It seems to be so. In the first place, while energy is not dependent upon motion as its cause, energy apparently cannot exist without begetting motion, and therefore, through motion, phenomena. It is the nature of energy to pass into motion. A world of energy which never begets motion,--is it really a world of energy? If it never begets motion, is it really energy? It seems a contradiction in terms. It is the nature of energy to beget motion. As for motion, it cannot exist without energy back of it. Neither can it take place without phenomena inevitably issuing from it. It can hardly be motion without being definite kinds of motion, and that means phenomena. And in turn phenomena of course cannot exist without motion, and back of the motion the energy, from which the phenomena issue. All of this is self-evident. Each of the three is inevitable with the others. None of the three can be without the others. No two can exist without the third.

By every test, matter does present itself as such an absolute threeness as we have been discussing in this comparison with the absolute threeness and oneness ascribed to the being of God by the New Testament.

Is matter then an equally absolute oneness? Is it a unity in that almost spiritual oneness in which "each one is the whole?" It seems impossible in the material world. But we can only ask, Are energy, motion and phenomena such a unity, in which each one is the whole? What now do we find?

When we approach the physical universe, we find that it consists entirely of phenomena. We have never seen, nor heard, nor felt, anything else of the material universe. Everything around us, of light, of colour, of heat, of pressure, of texture, of odour, of sound, is phenomena. We find, by our senses, or by instr

But we have found that all those phenomena are impact of motion of varied kinds upon our senses or upon instruments. They really consist of motion. We have found, therefore, that the material universe after all consists entirely of motion appearing to us as phenomena. This is not a contradiction. Anyone who knows modern science understands this.

And then we have found with equally absolute truth that the motion consists entirely of energy at work, and that therefore the material universe consists entirely of energy. That also is simple realism, to one who knows the facts of modern science.

Phenomena, motion, energy,--each one in turn is the whole. The impossible thing is simple fact. Matter, after all, is that kind of absolute oneness.

Well, there remains little to be said. Matter then is of the same extraordinary threeness and extraordinary oneness that is found in the mysterious and wholly spiritual Triunity presented by the Bible as the being of God.

Then are these three,--energy, motion and phenomena,--three modes of being? How do they compare at that point with absolute unity? Are they three things which the universe is?

This needs no long examination. Manifestly these three are three things which the physical universe is. We cannot say that the physical universe acts through energy and motion and phenomena, but itself exists apart from them. These are what it consists of. They are not primarily three things which it does. They are three things which it is. They are, then, as much as the three dimensions in space, or even as the three personal distinctions in the Triunity in the Bible, three modes of being.

Then come the questions of the relationships between these three elements in matter. Nothing could be a more definite triunity than space is. Space is simple, sheer triunity, and nothing else. But matter, which gives substance and reality to space, is richer in content and characteristics. Matter has, we know, a series of further relationships between its three elements, which are of the deepest interest.

The first relation is the logical order of the three elements.

Energy is first. That is self-evident.

Motion, which embodies energy, is second. That is equally clear.

Motion issues in phenomena, which are third.

This is the absolute, causal order in matter.

Energy is the source. It begets motion. It begets it perpetually. It embodies itself in motion. It works and acts through motion.

Motion makes energy operative in the physical world. It carries out, and executes, the possibilities of energy. Motion acts. Energy acts, but it acts through motion.

Phenomena proceed from motion. They do not embody motion, in the sense in which motion embodies energy. For they are, more accurately, the ways in which motion itself touches human beings. They proceed from energy through motion. They reveal and interpret motion.

All of this is self-evident to one who knows science to-day. None the less it is most deeply interesting to an intelligent person in the midst of the physical world. But the thing of vital interest just now is the comparison of these things with the elements in that mysterious Triunity presented to us by the Bible and Christianity as the being of God, who is the Ground of the physical universe.

In that remarkable Triunity the Father is represented as first,--the source,--the Son as second,-- the visible embodiment of the Father, and the Spirit, proceeding from the Son, as third, in a logical, causal order.

So, we have just seen, energy is first of the three elements in the triunity of Matter. It is the source in the threefold existence of matter. So also motion is second. It is the embodiment of energy. So again phenomena are third. They proceed from motion. It is an absolute, endless, unvarying logical and causal order.

The comparison up to this point is indeed very striking. We must surely go on with it.

In that mysterious Triunity presented to us as Father, Son and Spirit the Father, the Source, begets the Son, perpetually. He embodies Himself in the Son. He works and acts through the Son.

Just so modern science presents energy to us as the source in Matter. Energy begets motion. It begets it perpetually. Energy embodies itself in motion. It works and acts through motion. These are the self-evident words which we used to describe the relation of energy and motion a few moments ago.

In that other Triunity the Son is presented to us as the embodiment of the Father. He is the executive. He carries out and executes the plans and potentialities of the Father. The Son acts. The Father indeed acts, but He primarily acts through the Son, whether in creating, or in "becoming flesh," or in living, or dying, or rising, or judging.

So also motion is presented to us as the embodiment of energy. It is the executive in the physical universe. It carries out and executes the possibilities and potentialities of energy.

Motion acts. That is its very nature. Energy, of course, acts, but it acts, apparently, entirely through motion, whether in light, or heat, or electricity, or sound, or substance, or texture, or gravitation, or anything else which makes up the physical world.

So far the comparison produces an exact parallel. We must go on.

The Spirit, as that other Triunity is presented to us, "proceeds" from the Son. He does not embody the Son, in the sense in which the Son embodies the Father. He means not so much a concentration as a distribution of the activity of the Father through the Son. He is represented as a personal medium through whom the Son touches human beings. He reveals and interprets the Son. He is represented as proceeding from the Father through the Son.

So also phenomena are represented as proceeding from motion. They do not, as we have said already, embody motion in the sense in which motion embodies energy. They mean more a distribution than a concentration of the activity of energy through motion. They are the media by which motion touches us in the physical world, through light, sound, heat, hardness, softness, pressure, or other things which reach our senses. They reveal and interpret motion to us. And obviously they proceed from energy through motion. Like Space, Matter is a triunity, an absolute triunity, by all the highest tests of threeness and of oneness, and in this, and in all these other characteristics and relationships of which we have been speaking, it is extraordinarily,--exactly,--like the Triunity of the Bible.

How Does Time Compare?

How does the third great element in the structure of the physical universe compare with that principle of Triunity so strikingly found in the Divine Triunity of the Bible, and in space and matter? We know that this is a time-universe. We know that time is threefold, in its past, present and future. How do these three compare with the Triunity described in the New Testament?

Is Time an absolute threeness? Must it be no less than three, and no more than three? That seems to be self-evident. For time consists of three things. There are just three. They are past, present and future. There are always these three. There are only these three. There can be no more elements in Time than past, present and future. On the other hand, there can be no less. For no one of the three can be either of the others. It is essential threeness.

On the other hand, is time so essential a threeness that no two of the three can exist without the third? Clearly so. No one of the three can exist without the other two. No two of the three can exist without the third. For time cannot exist at all without all three. If there is no past, time has never existed until this instant, and a little later this instant also will never have existed. If there is no present, there is never any instant in which time exists. If there is no future, time ceases now, and indeed ceased long ago. Without any one of the three, time cannot exist. It is an absolute threeness.

Is time then an absolute oneness? Does it meet that really spiritual requirement of absolute unity? Are these three elements,--past, present and future,--so much one that each of them is the whole? It seems incredible that there could be yet another triunity in the physical world which should fulfill that almost impossible and spiritual condition. But does not each element in Time,--future, present and past,--include all of Time? All of Time is or has been future. The future includes it all. All of Time is or has been or will be present. The present includes it all. All of Time is or will be past. The past will include it all. At the beginning all Time is future. Between, all Time is present. At the end, all Time is past. Each one is the whole. They are as wholly one as that one is wholly three. It is an absolute triunity. Triunity could go no further.

Is this triune Time three modes of being? Manifestly. Past, present and future are three things which Time is, not three things which Time does. They are the essential nature of Time. Time is an essential triunity, as absolute, by all these tests which we have so far applied, as that simple, essential triunity which we call Space,--as absolute as that richer and fuller triunity which we call Matter,--as absolute as that kind of personal Triunity which the Bible describes.

Now Matter could be no more a triunity than Space is, as we have seen it. Nothing in the physical world could possibly be more triune than that. And Time could be no more a triunity than Matter or Space. Yet, because it is richer and fuller than Space, Matter has more relationships within itself. And we have found that these relationships are, as far as they go, entirely like those in the Divine Triunity described in the Bible. And Time, because it is, of these three primal things of the physical universe, the nearest to spirit, and is almost as much a thing of the spirit as it is of the material, has yet more relationships and characteristics within itself. Their analysis is of extraordinary interest, and will lead us into some untrodden realms, from which the mind may shrink, but which it will accept as reality.

The Relationships of Time

How does time exist? What is its source?

Here is where we must diverge from those who have heretofore discussed the nature of time. We cannot safely so diverge from all who have gone before unless the thing which we discuss can be shown to be self-evident. That is what we must show.

How does time exist? What is its source? Not the past. Carelessly we think of it so, as coming out of the past. Moralists, poets and scientists speak of it so. We speak of the stream of time as flowing out of the past. Even the upholders of the new Science, which makes much of time as a basic reality, along with space, of the physical universe, and strongly emphasizes the invariability of the movement of time, describe that movement as proceeding out of the past through the present into the future.

But time does not come out of the past! It comes out of the future. And it does not flow into the future! It flows into the past. This may bring a shock to one's habit of thinking. It has brought a shock to some who have discussed it as it is presented in these pages. We have never thought of it in that way. None the less it is the self-evident fact. We have but to take a definite date, a definite piece of time, and trace its course down the stream of time, to find at once whether that section of time moves from past through present into future, or from future through present into past. Consider, for instance, that section of time which we call "to-day," the day in which you read this page. For a long time this day was "next year," far in the future. Then it was "next month," still in the future. Then it was "next week," in the near future. Then it was "to-morrow," in the immediate future. Then it became "today," in the present. Soon it will be "yesterday," in the immediate past. Then it will be "two days ago," in the near past. Then it will be "last week," in the recent past. Then it will be "last month," in the receding past. Then it will be "last year," far in the past. Manifestly, that section of time which we call "to-day" comes out of the distant future, first into the near future, then into the present, then goes into the recent past, then disappears into the distant past. That is the unbroken order of the motion of time. That is its invariable direction. Never does it flow the other way, from past to future. Never does yesterday turn back in its flight and become to-day, or to-day become to-morrow. Never does the past pass into the present, or the present into the future. No. It is the other way. To-morrow becomes to-day. To-day becomes yesterday. The future becomes the present. The present becomes the past. The future is the source, it is the reservoir of time which will some day be present, and then past. The present is the narrow strait, it is the living instant, it is the flashing reality, through which the vast oncoming future flows into the endless receding past.

Why then do we usually think of time as coming out of the past? What is the reason for this common fallacy? The answer is simple. We get the impression that time comes into the present out of the past, because the human race and human history come into the present out of the past. The human race passes from past through present into future. Therefore we have fallen into the habit of thinking that time follows the same order. But it is not so. Time goes the other way. The human race comes to us out of the past, and time comes to us out of the future. We do not go with time. We continually meet it, instant by instant. That is why the present is always instantaneous, because we do not go with it, but constantly meet it, moving from past to future while time proceeds from future to past. This is the procession of time. The future is the reservoir out of which the present comes. The future is the source.

The Future is the source. The Future is unseen, unknown, except as it continually embodies itself and makes itself visible in the Present. The Present is what we see, and hear, and know. It is ceaselessly embodying the Future, day by day, hour by hour, moment by moment. It is perpetually revealing the Future, hitherto invisible.

The Future is logically first, but not chronologically. For the Present exists as long as Time exists, and was in the absolute beginning of Time. The Present has existed as long as Time has existed. Time acts through and in the Present. It makes itself visible only in the Present. The Future acts, and reveals itself, through the Present. It is through the Present that Time, that the Future, enters into union with human life. Time and humanity meet and unite in the Present. It is in the Present that Time, that the Future, becomes a part of human life, and so is born and lives and dies in human life.

The Past in turn comes from the Present. We cannot say that it embodies the Present. On the contrary Time in issuing from the Present into the Past becomes invisible again. The Past does not embody the Present. Rather it proceeds silently, endlessly, invisibly from it.

But the Present is not the source of the Past which proceeds from it. The Future is the source of both the Present and the Past. The Past issues in endless, invisible procession from the Present, but, back of that, from the Future out of which the Present comes.

The Past issues, it proceeds, from the Future, through the Present.

The Present therefore comes out from the invisible Future. The Present perpetually and ever-newly embodies Lie Future in visible, audible, livable form; and returns again into invisible Time in the Past.

The Past acts invisibly. It continually influences us with regard to the Present. It casts light upon the Present. That is its great function. It helps us to live in the Present which we know, and with reference to the Future which we expect to see.

All of this is indeed remarkable. But here is something yet more remarkable, which we can discover for ourselves by a simple experiment. It is possible to take the preceding paragraph, with its carefully detailed and self-evident setting-forth of what we all know to be the relations of Future, Present and Past, and to read it actually without change, substituting the word God for the word Time, the word Father for Future, the word Son for Present, and the word Spirit for Past, and have a detailed and exact description of the relations between Father, Son and Spirit as the Scriptures present them.

Let us try it. "The Father is the source. The Father is unseen except as He continually embodies Himself and makes Himself visible in the Son. The Son is what we see, and hear, and know. He is ceaselessly embodying the Father, day by day, hour by hour, moment by moment. He is perpetually revealing the Father, hitherto invisible.

"The Father is logically first, but not chronologically. For the Son exists as long as God exists, and was in the absolute beginning of God. The Son has existed as long as God has existed.

"God acts through and in the Son. He makes Himself visible only in the Son. The Father acts, and reveals Himself, through the Son. It is through the Son that God, that the Father, enters into union with human life. God and humanity meet and unite in the Son. It is in the Son that God, that the Father, becomes a part of human life, and so is born and lives and dies in human life.

"The Spirit in turn comes from the Son. We cannot say that He embodies the Son. On the contrary

God in issuing from the Son into the Spirit becomes invisible again. The Spirit does not embody the Son. Rather He proceeds silently, endlessly, invisibly from Him.

"But the Son is not the source of the Spirit who proceeds from Him. The Father is the source of both the Son and the Spirit. The Spirit issues in endless, invisible procession from the Son, but, back of that, from the Father out of whom the Son comes.

"The Spirit issues, He proceeds, from the Father, through the Son.

"The Son therefore comes out from the invisible Father. The Son perpetually and ever-newly embodies the Father in visible, audible, livable form; and returns into invisible God in the Spirit.

"The Spirit acts invisibly. He continually influences us with regard to the Son. He casts light upon the Son. That is His great function. He helps us to live in the Son whom we know, and with reference to the Father whom we expect to see."

This is more than likeness! It is identity, not of substance, but of principle. "Analogy," unless in the deepest mathematical sense, falls far short of describing such repetition of every detail and every intricate, mysterious relationship. It is clearly the same principle, seen first in terms of God in the Bible and then in terms of Time in the world. This universe in which we live is a time-universe. All thinkers recognize that. Many scientific thinkers lay great emphasis upon it today. And as a time-universe it is, as a matter of simple fact, whatever our conclusions from it may be, exactly like that Triunity of Father, Son and Holy Spirit which the Scriptures called the Bible so fully describe.

The Structure of the Universe

Out of our study of Space and Matter and Time emerges truly an extraordinary thing. We set out to discover whether we had before us a vast structure of things, including on the one hand space, matter and time, which are the fabric of the physical universe, and on the other hand God, who is the ground of space and matter and time. There can be no question as to what we have found. We have found a vast triune structure of the space-matter-time universe. It is truly a triuniverse. It is, in terms of space, matter and time, exactly like the Triunity presented by the Bible as the being of God. The likeness is so exact that it is clearly the same principle, first in the being of God, and then in the structure of the physical universe. It is not analogy. For it is much too exact for that, unless in the mathematical sense of analogy. For it presents much more than similarity. When two geometric figures cover exactly the same points, with exactly the same lines running between the points, they are the same figure, even though the two are in different places, of different sizes, and of different material. And when one of them faces the other in a reflecting mirror, they are clearly the same figure, the one reflected from the other. So these two, the Triunity of God in the Bible, and the triunity of the physical world which reflects God, are in that sense the same triunity. I do not think that anyone will misunderstand this. There is no intention of declaring that these two triunities, face to face, and so extraordinarily alike, are identical in the sense that the physical world and God are one. They are manifestly not the same triunity in the sense that they are the same substance. One is impersonal. The other is Personal. There is no trace of "three centres of personal consciousness" in the three dimensions of space, or in energy, motion and phenomena, or in future, present and past. Clearly the two are not the same substance. But they present the same triune structure, at every point, expressed in the one case in terms of Divine personality, in the other in terms of space, matter and time. They are the same, just as the image of yourself in the mirror is more than similar to you, it is exactly the same, in form, colour and movement, but in terms not of flesh and blood and spirit, but of glass and quicksilver and light. So it is manifestly at every point the same triune structure, in the one case presented in the being of God, in the other in the universe which reflects God.

All of this converges upon two great conclusions. Their value lies in the fact that they are self-evident. For this reason we will put them as simply as possible.

1. The Triunity shown in the Bible manifestly presents a vast and adequate reason for the triune structure of the physical universe. For the reason ought to be in God. The universe ought to reflect God, its Maker and Ground. That should be the reason for the general character of the universe. The structure of the universe ought to reflect the structure or being of God. Any theist will agree with this. Such Triunity of Father, Son and Holy Spirit in God presents therefore an adequate original and reason for the exactly similar triunity in the fabric of space, matter and time. Whether one accepts that Triunity or not, one must admit that, in view of the exact likeness, it does present an adequate original for the universal triunity. It gives as a reason for the universal triunity simply this, that the universe mirrors its Creator. It means that the universe is essentially like its God. It declares that the creation reflects the Creator.

2. The fabric of space, matter and time presents a universal and exact confirmation of that Triunity in God. For the one vital and conclusive proof which the physical universe can give of that Triunity is that the universe should reflect it. This means that such triunity should be found as basic in the universe of which God is the Ground, and which in other ways reflects Him. And here is such triunity everywhere in the universe. It constitutes the very structure of the universe. It is at every point exactly such triunity in terms of space, of matter and of time as it must be to reflect that Triunity of Father, Son and Holy Spirit in God. It is exactly the confirmation which the universe ought to give. This is, of course, the method of science. "If such a hypothesis is true, such an effect will be found in the physical world. It is found. The facts then confirm the hypothesis, and establish the law." So much is this the modern method, so decisive it is for us, that when infinitesimal deflection was found in rays of light passing near the sun, the entire theory of Relativity, to its remotest branches, was regarded by many scientists as fully confirmed. Here is such confirmation of the Triunity of God in the facts of the physical universe. But the facts consist not of minute and disputable data, but of the whole self-evident structure of the space-matter-time universe. And their bearing upon the case is not subtle and abstruse, but lies in their exact reproduction, at every point in space, matter and time, of every possible point in the Triunity of God presented in the Bible. It means that the universe reflects its God, and thereby attests that Biblical Triunity of Father, Son and Holy Spirit, which it so completely reflects.

It means, too, that in this general triunity the physical universe brings to us its strongest proof of the existence of God. The Bible presents God as all-wise. It describes Him as all-powerful. It depicts Him as holy. It shows Him as loving. And not only the mind of man responds to this. The physical world, also, seems to many to reveal something of these things. All-wise? Yes, the world shows that, as far as your eye or the microscope can penetrate. All-powerful? Yes, that can be found anywhere, and as far as you or the telescope can see. Holy? That seems to be embodied in the laws of nature. Loving? That seems to be reflected in part in the loveliness and the comforts of nature. But now comes a far more drastic test. The Bible depicts God, in His very being, as Three in One, with a marvelous and intricate group of characteristics and relationships. Can the physical world reflect even this, so elaborate, so complete, not vague at all, like a general characteristic, but precise, many-sided, articulated? Yes, as we have been seeing, the physical world parallels that Triunity, and all its relationships and all its functions, with an accuracy, a fulness of reflection and a precision which are impossible in any other thing about God. By this His existence may be recognized, with supreme clearness, from His world, whether we call it His reflection presenting Him face to face with the world, or His creative impress left upon the world, or His visible vesture revealing His moving presence in the world.

It is your great privilege as a thinker who deals with facts to have if you will a basic vision of the universe. It is a great thing to have that vision. It brings before one a transformed and illumined universe. No one can really know the world of Space and Sense and Time who fails to recognize this universal Fact of Divine Triunity. "Within its depths," said Dante of that central Trinity, at the climax of his vision of the universe, "within its depths I saw ingathered, bound by love in one volume, the scattered leaves of all the universe; substance and accidents, and all their relations, as though together fused, after such fashion that what I tell is of one simple flame." To see Him invisibly and yet visibly everywhere in triune Space! To recognize the mighty working of the Triune Master of the world in energy, motion and phenomena! To feel in the endless generation and procession of Time the presence of Him who is the Father, the begotten Son and the proceeding Spirit!

When you look up at the stars, as you often do, and see in them a million burning mirrors of the mind and will of God, remember that around them, through them, beyond them, is the greater, more absolute, complete reflection in Space itself, unseen, silent, everywhere, infinite, of His very Being, Father, Son and Holy Spirit, the one God. And then remember that this moving, shining universe of Matter around you, with its balance of the Divine mind, its urgency of the Divine will, and its beauty of Divine love, is in its entire nature and structure of energy, motion and phenomena the absolute and complete reflection of the Divine Being itself, Father, Son and Holy Spirit, Three in One. And then remember that those vast pulsations and measures of Time, beating through the stellar universe, with the ancient past, the present and the distant future all visibly before you, are the complete and perfect reproduction of the Three in One who made and upholds this universe of wondrous worlds.

II

THE INNER UNIVERSE

The inner universe.--Human existence--Why is it what it is?--Famous failures to meet the tests--Self-realization, or how we recognize ourselves, and what it reflects--Self-determination, or how we decide things, and what it reflects--"Personal existence itself"--The marvelous series of facts--Even three personal centres of consciousness--"The whole of personal life"--"A perfect human likeness"--"In all men"--The method of science.

THERE is a more wonderful universe than that which we know as the outer world. It is the inner universe. It is the world within our own souls. You may call it human life, or personal being, or humanity. It is the universe of human existence.

Why is human existence what it is? Is there some great reason for its characteristics and its structure? Why is man constituted exactly as he is? If he grew to be what he is, why was he so constituted as to grow into exactly what he is?

Of what then is human existence constituted? Of what does it consist? Can we see it in a broad and simple, a self-evident way?

The answer to this question cannot be found by plunging into subtleties of metaphysics or psychology. Those are not broad and simple. Nor are speculations, no matter how brilliant, to be classed among self-evident things. The answer is to be gotten rather by standing off and looking at human life in its most everyday aspects. There we shall see the broader, simpler structure of human life. Of what does human life as we know it around us, in our friends, in people everywhere,--of what does it consist?

Well, first of all, and very obviously, it consists of persons. It is made up of people. Human life, personal life, consists entirely of people, of persons.

Think then of them, of persons. Here is an individual whom you know. In knowing him, you know, as distinguished from all other kinds of things, a person. He is a human being. He is a person.

That, you say, is very simple. So it is. But that is what we want. It is simple, and broad too, and obvious, and fundamental. Here then is an individual whom you know. He is a person.

But that is not all, except in the very shallowest thinking, or the most thoughtless acquaintance with human beings. Suppose that you come to know this person well. You talk with him. You become intimate with him. He tells you his thoughts and feelings. You learn to know why he acts and speaks as he does. You understand his sources of action. Now you say that you know him so well that you know his very nature. We all understand this. You know all of that in him which is back of the person whom most people casually meet and know. You know his nature. That is another great thing, then, about personal existence. You know his nature. You know his inmost primary self.

Now another question, a very simple analysis, not at all abstract, but a matter of daily life. How do you know this person, and this nature back of the person? The answer is evident. You know him as he touches you, as his life comes in contact with your life, and influences you, and impresses you. You know him, you say, by his personality. We understand clearly what we mean by that. We lay great emphasis upon it in modern life. His personality is he, the person, as he touches the lives of other people. It is the only way in which you can know him. It is the person as he reaches out and influences and impresses you and others. That is a third great thing about personal existence.

Let us gather up what we know. This is what every human being is. A person. Back of the person, a nature. In the lives of others, a personality.

These three things are surely simple and broad. They may never have been gathered together in this way before, it is true, but they are the self-evident elements of personal life. They sum up all that you yourself are, or that anyone else is. They are the three things which every human being is.

Why is it what it is?

Why is human existence just such as this? Why is it in its simple outline exactly what it is? Is there a reason for this threefold structure of human personal life? It cannot be so without some basic reason for it.

Naturally, in view of all that has shown itself to us in the physical universe, one may ask, "Is the reason for this threefold structure of human existence to be found in the universal triunity of Space, and of Matter, and of Time, and in the threefold being of God as presented in the Bible?

Is man included in a vast triunity of all things? Is this the explanation of the threefold being of Man, that it reflects God?"

That is not at all to be taken for granted. One cannot jump at so great a conclusion. But it is reasonable to ask the question.

For man is a vital part of the universe. He is bound up in bodily life with space, for his body has dimensions, and with matter, for his body is composed of that, and with time, for his physical existence is a time existence. And he is bound up in soul with those three, also, for he must think spatially, and his soul goes in and out through the doors of the senses, and his mind lives a strictly time existence.

And the being of man does reflect God. Man is much more like God than the physical universe is. That is why we can understand God. The inner world may not be greater than its twin universe, the outer world. For the physical universe seems to be infinite, and the mind is not. At least we know that the physical universe is greater than the mind can comprehend. But the mind is higher, for it can understand itself and this physical universe and God. And it understands God because it is like Him in personal being. In this it is far higher than the physical universe. Does this inner universe of personal existence, then, as we know it in man, agree with the triunity of the outer world and of the God of the Bible, and make that triunity truly universal? Is that why man is what he is?

How Man Reflects God

We know that the being of man reflects God. We know this in a simple but profound way. The things shown to us in the Bible and in Nature about God are confirmed to us by likenesses of them in man. We compare them with their reflections in the mirror of human existence.

We know for instance what we mean by an omniscient God, a God who knows all things at once, and knows them not simply by reasoning, but by direct vision of all things at once. For we have a faint reflection of that, in our gift of memory, which releases us from seeing only the things of the present, and sees things of other days as though they were to-day. And we, also, have intuitive knowledge, of things back of reasoning. We know our own existence, and God's existence, and mathematical axioms, and that effects follow causes. In our memory and in our intuition we have enough reflection of Divine omniscience to let us comprehend what that omniscience means.

We know what an all-powerful God means. The idea may involve some problems to us, but the essential fact presents no difficulty to our minds. We understand instinctively the supreme, overcoming power of a Divine will. For we know something of it, because of the marvelous power of our human will, which overcomes obstacles, accomplishes the impossible, melts hindrances into channels of power, and carries circumstances before it.

We understand even a God who is omnipresent, everywhere at once,--not as an atmosphere is, a part of it here, a part of it there,--but in Personal presence,--He, the whole of Him, here,--He, the whole of Him, there. That is a thing which may truly seem to be without parallel in us. But we have a marvelous faculty, as mysterious in its limited human way as that Divine omnipresence. I sit here at my desk, but this instant I am in London, or in New York, or in the country, or in India, instantly, absolutely,--seeing, thinking and feeling not here but there. In that mysterious and vivid power of transporting myself and my whole consciousness to any distant place, in an instant, so that my mind is there, and is not here,--in that which we call imagination,--we have a reflection of an omnipresent God.

And a holy God. A sinless God. How can we understand that in God? But, though we are all of us sinful, we do understand His sinlessness, His absolute holiness. We understand it when we are told of it. We could even reckon it without being told of it. For we have a marvelous reflection of it in us, which tells us what holiness means, and leads us to know instinctively that holiness is what personal being at its highest ought to be. We have conscience,--which makes us moral beings like Him,--conscience, the image and echo in us of the holiness of God, telling us unmistakably what is holy and what is not,--what is right and what is wrong.

As for the love of God, how easily we understand that, because we also love.

Although broken and defaced, although dim and dark, man is the mirror of God, and instinctively understands what God is like. Though God is infinite and we are finite, though He is holy and we are sinful, yet we can easily and instinctively understand what He is like, because He is like us, and we are like Him.

Does man then, with a body of space and matter and time, and a mind conditioned also by those things, and in his personal being reflecting all the characteristics of God as shown in the Bible,--does man agree in his being with the structure of the universe, and with

what the Bible presents as God's very being, that He is Three in One? These things of which we have been speaking are characteristics of God and of man. But what about the being of man? Does that reflect God? Is that why it is person, nature and personality?

Whether a reflection of God in man is the reason for the structure of human existence depends on whether this threefold being of man in person, nature, and personality is so exactly, or, if you will, so absolutely, like the Triune being of God as found in the Bible, as to be obviously the same principle of being in the Creator and the created. The being of man, if it reflects the being of God, ought to be not less exactly like Him than the physical universe is. It ought to resemble Him even more closely. Is that the case?

The requirements in discovering the answer to that question are not uncertain.

Genuine candor is one,--the readiness to take facts as we find them.

Precision is another,--a care as great as in any experiment in physical science,--accuracy as complete as in any science of the mind,--mathematics, or logic, or sane metaphysics,--even though it means repetition.

And reality. One should avoid fancies, or ingenuities, or poetic analogies. Ordinary analogies should be ruled out. They have their value as illustrations, especially in public speech. But they prove nothing. The facts should be self-evident. They ought to be in the basis of personal being. They ought to be living facts, and evident in daily life. And they ought to be so genuinely and exactly like the mysterious distinctions and relations in the Biblical Triunity of Father, Son and Spirit,--and therefore like the triunity in the space-matter-time universe, that anyone of open mind can see that they are the same thing, expressed in terms of human life.

For there are certain famous comparisons between the being of man and the Divine Triunity which fall short very surprisingly when one comes to such analysis. They seem to be based on the nature of personal existence in man and so to be arguments for "necessity" in the personal existence of God. Such arguments for something as necessary in the being of God need to be approached with great care. To argue that God must be so and so because man is so and so is likely to be a futile business. We do not know why God is what He is. Surely He is not what He is because of what man is. That is the trouble with arguments that God must be so-and-so because man is. For man and his being are not the cause of God and His Being.

Witness the argument from Divine Love:--"God is love. Love must have an object. Eternal love must have an eternal object. That means at least one other person within God."

This implies more than one in God. It emphasizes the richness of God's inward nature and experience. But it does not prove anything from necessity of being in God or man.

For love is not an abstract necessity of being in God or man. "God is love" comes from the Bible, which also tells us that God is Father, Son and Holy Spirit. The Trinity could equally well be used to prove that God must be Love.

Nor is there the basis of such an argument in any reflection in man's being. Love within the circle of one'son being in man is self-love, which is the opposite of God's love.

Witness the argument from the nature of God as Father. "God is by eternal nature 'Father.' But there cannot be a Father without a child. Therefore the eternal Sonship is necessary to the eternal Fatherhood."

But this is not an argument from necessity. It is an argument from the Bible. It is the Bible which depicts God as Father. The Biblical depiction of God as Father, Son and Holy Spirit is equally valid, and much more direct as an argument for the Trinity in God.

As for a human reflection, there is nothing in man which reflects Fatherhood and Sonship within one being.

The trouble is that arguments from the being of man to prove that God must be similar fall far short of their goal. "For man and his being are not the cause of God and His Being." They do not determine the nature of God. God is not a reflection of man.

If any of these arguments presented something in man which exactly resembles in a finite way the Triunity of Father, Son and Holy Spirit in the Bible, they would have the force, not of arguments from necessity, but of confirmatory proof, by such reflection in man, of the Trinity which the Bible presents.

Witness the argument from the older psychology, of intellect, affections and will, or mind, heart and will. It sounds now, but it has appealed to many in the past.

But it is not certain that this is a necessary threeness, in God or in man. Where does memory belong? Or imagination? Or conscience?

Neither are intellect, affections and will in any real way like the Three in the Trinity of the Bible. Although Father, Son and Spirit are spiritual, intellect, affections and will do not at all reflect them as Matter does, or Time, in the physical world.

We must turn away from every argument for "necessity" in the being of God, based upon the being of man. They fall short. They do not reach the being of God. "Man and his being are not the cause of God and His Being."

Nor are love, or fatherhood, or mind, heart and will, in man so much like the Three in God in the Bible as to be a reflection of any such Divine Triunity.

If there is a genuine likeness of Father, Son and Holy Spirit in man, it will not prove that God must be so, or show why He is so. Man is not the reason for God's nature. God is the Cause, and man is the creation. God is the Original, of which man is the reflection. But as a reflection, such a likeness in man will be a vivid evidence of the Original.

Self-realization

It has long been known that the act of self-realization, of realizing one's own existence, shows a real, though dim and limited, resemblance to the three distinctions in the Triune Being of God in the Bible.

A person, we say, is one who can say "I." It is one who realizes himself. We do not say "I" unless we realize that we exist, unless we see our own existence.

There are three factors in this process.

1. I, who realize or see. That is absolutely and evidently the first factor.

2. That self whom I see or realize. That is the object of the act of such self-realization.

But one does not realize oneself or one's existence unless one goes further than this. If I see myself, but do not realize that it is myself, if I see my existence, but do not realize that it is my existence, it is not self-realization, and I do not say "I." The kitten chases its tail, and does not recognize it as a part of itself. The baby puts its foot into its mouth, not knowing what it bites. The little child never says "I," but says "Johnny is tired," or "Mary wants a drink," and does not say "I" until self-realization begins. The insane person sees self but thinks it is some one or something else, Napoleon, or Caesar, or a steam-engine. All these are examples of a process of self-realization which is incomplete because it stops short of the third element. There is, therefore:--

3. Myself recognized as myself. Then there is unity, and in the unity there is realization of self. The first, "I who see," and the second, "I who am seen by myself," are united by the completion of the circle in the third, "I who am recognized by myself."

These three somewhat dim distinctions in the process of personal self-realization in several ways reflect the distinctions and relations between the Three Persons of the Trinity in the Bible. There are of course limits to this likeness. These are due largely to the very limited characteristics of these three distinctions in self-realization. And because they are dim and abstract these distinctions cannot be said to be real to most people. They can hardly, for instance, be called three centres of consciousness, as any likeness of the Divine Triunity in man must be, when no one is actually conscious of them in daily life. They are not as real in experience as the Trinity is to Christians! Something must make them more real and conscious before we can call them three centres of consciousness, or see them as a vivid reflection of the Divine Triunity of the Bible.

Self-direction

There is a more distinct human likeness of centres of personal consciousness. It has appealed to many. It occurs in that daily experience of yours in which you consider what you will do; and look at the problem from this point of view, and then from that point of view, and then from the other point of view,--and then decide.

A business man is faced with a question of finance. "Shall I go into this business opportunity? As a financial man, I think I will. It is a sound financial proposition. The returns will be large and steady, and long-continued."--"But there is another point of view. As a husband, I am not sure. The business will take me much away from home, altogether too much. My wife and I may even grow somewhat apart. As a husband, I think I will not do it."--"And now there is another point of view. Let me consider the question from that point of view. I am a father. If I go into this business, while I shall make money for my children, I shall neglect them personally, just in these years when they and I ought to know each other best. No, I think I will not do it."--"But there is another point of view. I must remember that I am a citizen. This business will give me power, and financial influence. I can use these as a citizen. I can do great good in the community. I can have the strong influence a man ought to have. I don't know,--perhaps I will enter this business." "But there is still another point of view. I am a Christian. How does that affect this business opportunity? Why didn't I think of that before! I shall have to do some things in it which I cannot do as a Christian. It is a respectable business, but to succeed in it I shall have to do some things which I cannot do as a Christian. Really, I can't do it." And then, after all these personal points of view, as a business man, as a husband, as a father, as a citizen, and as a Christian, he, the human being, all of these things in one, decides: "I won't do it."

This power to take different points of view, to have all one's consciousness of oneself centered first in one point of view, and then in another, and then another, and then after such division and debate, finally, one's unified self, all these points of view in one, to decide,--this truly presents something like personal centres of consciousness in one human being. But they are not three. Nor are they in any way like Father, Son and Holy Spirit in God as presented in the New Testament. Nor are they invariable; they are constantly changing into many varied points of view. They are distinct, and real in daily experience, but unless something makes them three, and invariably the same three, and like the Three in the God of the Bible, they cannot be called a finite reflection of the Divine Triunity of Scripture, such as we find so marvelously in Space, in Matter and in Time.

Personal Existence Itself

None of these things are personal existence. They are aspects or activities. They are characteristics of human existence. They are not existence itself.

The being of man, the broad, simple, self-evident structure of human existence itself,--that is what we must deal with.

We have already seen that realistic pattern and structure of personal existence, and seen it as Person, Nature and Personality. Some have objected that this triunity in man is new, and not already known in psychology or philosophy. But when all familiar likenesses of the Trinity fail, it would seem that newness is an advantage. In the Outer Universe new vision has not proved a disadvantage.

Then let us ask in regard to that new but very real pattern of human life, in the person, and his inner nature, and his personality affecting others:--"Is this what it is because of what God is? Is man what he is because in his threefold person, nature and personality he reflects his Maker?"

This is a matter for careful comparison of those Persons in God as presented by Jesus and the New Testament with these distinctions found in human life. And must there not be accuracy at least as great as in any important experiment in physics or astronomy?

In the Triunity of God in the Bible there is Absolute Threeness. There are never any more than three presented. There are never less than Three.

The distinctions of nature, person and personality in man are in a finite way an absolute threeness. "The person",--"back of the person a nature",--"merging in the lives of others, a personality",--there are no other such distinctions in daily actual human life. There are no more than these three. Can you think of any other such factors? There are no more. And there are no less. For one is not a personal being unless all three elements are present.

In the Triunity of God in the Bible each of the three, Father, Son and Holy Spirit, is presented as so distinct that no one of the Three can be either of the others. There can be no less than the three.

Each one of the three factors in man is so distinct that no one of the three can be either one of the others. There can be no less than the three.

And in absolute Divine Threeness, such as that of God in the Bible, each of the Three is inevitable to the existence of the others.

In the triunity in man's being each of the three factors is inevitable to the others. Without the person there is no nature or personality. Without the nature there is no person, no personality. And there is no person, no nature, without personality as a consequence.

In the triunity of God in the Bible there is absolute Oneness. The Three are so much One that each one is actually the whole. Each one is not simply a part of God. Each one is the whole. Each is God.

Person, nature and personality in man are also one. They are so much one that each is the whole man. Each one is not simply a part of the man. Each one is the whole.

The person is of course the whole man. His nature is the whole man, also. His nature is really all of him, all that he actually is, working together to be the source of what he, the person, says and does.

And his personality,--that is the whole man, too, as he affects others. He may try for an artificial personality, based on only a part of himself and his nature. But the totality of what he is will none the less, in spite of himself, and perhaps unconsciously to others, form his personality as it affects others.

Each of these three, then, person, nature and personality, is truly the whole. By that highest test the three are deeply and entirely one.

In the Trinity of God in the Bible the Three, Father, Son and Holy Spirit, are what God is, not simply what he does, or three ways in which he acts. They are three modes of Being.

The three factors in personal human life, the person, the nature, the personality,--are manifestly three things which the man is. They are not three things that he does. They are not mainly three ways in which he acts. When he is not acting he is these three. The person is what one is. One's nature is what one is. One's personality is what one is. They are three modes of being.

In all of these tests the simple, realistic, self-evident structure of human life is seen to be an exact human likeness of the Distinctions in the Divine Triunity of the New Testament. But there are other tests. There are all those relations and those characteristics described in the Three in One of the Bible. Is there any likeness to these also in human existence? Can such a mysterious, infinite Triunity as that be paralleled or imaged in a finite human life? This comparison cannot be too careful. We want no poetic analogies or metaphysical abstractions. This is a time for exactness, a laboratory experiment in the reality of human life. We may be pardoned if we make the experiment one of extreme precision. We need not let this weary us, if we remember that it is the way to certainty, and if we let ourselves feel the thrill of such a series of facts as now unfold themselves before us.

The Marvelous Series of Facts

In God as found in the Bible the Father is the source. The Father is unseen. He reveals Himself in the Son.

In man the nature is the source. The nature is unseen. It reveals itself in the person.

In that Triune God the Son is the visible embodiment of the Father. He is begotten, eternally, from the Father.

In this triune man the person is the visible embodiment of the nature. The person is begotten continually from the nature.

The Father reveals Himself in the Son. No man hath seen God. The only-begotten Son, he hath revealed Him. He that hath seen the Son hath seen the Father.

The nature reveals itself in the person. No one has seen the nature by itself. The person, begotten from the nature, reveals it. He who has really seen and known the person has seen the nature.

The Son is the visible embodiment of the whole Godhead, including both Father and Spirit. In him dwelleth all the fulness of the Godhead bodily.

The person is the visible embodiment of the whole being, including both nature and personality. In the person, then, dwells the whole being bodily.

In God in the Bible it is the Son who especially acts. He is the executive.

In man in finite life it is the person who especially acts. He is the executive.

The Son works in certain ways, in other lives, by the Spirit. Yet that is also the Son himself working in other lives. The Spirit is his other self.

The person works in certain ways, in other lives, by his personality. Yet that is also the person himself working in other lives. His personality is his other and unseen self in other lives.

The Spirit proceeds, goes out, from the Son. He is not the embodiment of the Son. Quite the contrary. The Spirit is invisible. He is never seen, but He is felt. He works unseen, in other beings. He reveals the Son.

The personality proceeds, goes out, from the person. It is not the embodiment of the person. Quite the contrary. The personality is invisible. It is never seen, but it is felt. It works unseen, in other beings. It reveals the person.

The Spirit proceeds not only from the Son, but from the Father. He proceeds from the Father, through the Son. That is the clear presentation found in the Bible.

The personality proceeds not only from the person, it proceeds from the nature. (There may be an attempt to form an artificial personality, by a false cordiality, a false forcefulness, a false sincerity. But it does not really succeed. The personality, proceeding from the person, proceeds ultimately from the inner and unseen nature of the person.) It proceeds from the nature, through the person. That is the clear presentation found in human life

The Son sends the Spirit. He sends the Spirit, he tells us, from the Father.

The person sends out his personality. He sends out his personality, as we know, from his inner nature.

The Father sends the Spirit, in the name of the Son, we are told.

The nature sends out the personality, with the name of the person upon it.

The Spirit reveals the Son, but He also just as truly reveals the Father, in and through the Son.

The personality reveals the person, but it also just as truly reveals the nature, in and through the person.

In all of this in God there is a logical, causal order. The Father is first. He is the source of all that God is. The Son is second, embodying and perpetually begotten from the Father. The Spirit is third, proceeding from the Father through the Son. It is not that one is first, one second, one third, in deity, for all are God; nor that one is first, one second, one last, in time, for all are eternal. It is a logical, causal order.

In all of this in man there is a logical, causal order. Nature is first. It is the source of all that you are. Person is second, embodying and perpetually begotten from the nature. Personality is third, proceeding from the nature through the person. It is not that one is first, one second, one third, in identity, for all are you; nor that one is first, one second, one last, in time, for as early and as long as you exist all three exist. It is a logical, causal order.

This surely is absolute likeness! It goes far beyond the likeness between the triunities of space, matter and time and the Biblical Triunity of God. Where are there any parallels like it? It breaks through the realm of analogies, and becomes identity. We gaze on the same principle, first in terms of Divine life, and then in terms of human life. It is identity, not of substance, but of pattern and structure, extending into every possible detail. As we said in talking of the physical universe and the Biblical Triunity, we can say now of this likeness between the Triunity of God in the Bible and the triunity of man,--"They are the same triunity, as the image of yourself in the mirror is more than similar, it is yourself, but in terms not of flesh and blood and spirit, but of glass and quicksilver and light." So it is the same triune structure, in the one case presented in the being of God, in the other in man.

Even Three Personal Centres of Consciousness?

The likeness is so extraordinary,--can we then go on to a further test? It was a test impossible for the physical universe. But nature, person and personality lie in the realm of personal existence. Does this extraordinary reflection in man extend its likeness of absolute Triunity to the point of reflecting three Personal centres of consciousness in one Being?

We have found nothing like that. Space has nothing of that kind. Matter and time say "It is not in us." Nor have we seen any such thing in man. "I who see," "I who am seen" and "I who am realized" in self-consciousness are not conscious enough to be three personal centres of consciousness. The points of view in inward debate and decision are not apparently three, but endless. Do Nature, Person and Personality do what self-realization and self-direction do not? Do they present three personal centres of consciousness in one being? Do they show to our astonished vision a thing as marvelous as that?

We have, as we have said, various centres of personal consciousness in inward debate and decision. We experience them every day. They are intensely real. But they are not three, unless they group themselves in some inevitable way in three habitual, invariable, all-inclusive centres of consciousness. Do they do this? Are they naturally three? How do they group themselves?

We should find the answer, if we can find it at all, in the light of our new three factors of Nature, Person and Personality, the structure of every day human existence. What do these three factors show us?

First, there is the simple, elementary point-of-view or centre-of-consciousness. It is the point from which we think and see first of all. It is the simple, personal point of view,--of me, the person, simply as self, and of my proposed action as affecting myself and my so-called personal interests. "Will it be good for me?" "How will it affect me?" "Will it benefit me?" That is the first habitual and universal point of view. There is no wrong in it, if it is not the only point of view.

Second, there is the point of view of my essential, true nature, and of my proposed action as true or false to that nature. "Is it true to my higher nature?" "Is it true to my real self?" That is the second great, inclusive point of view.

Third, there is the point of view of me as related to others, and as my action will affect others,--family, community, and so forth. "Will it be good for others?" "Will it make others happy?" "Will it do harm to others?" That is the third great point of view.

All my consciousness in thinking and in deciding centres itself constantly in these three points of view in turn. They include all the possible centres of my consciousness. All of the changing points of view in my swiftly moving consciousness come into these inclusive personal centres,--myself as a person,--my nature,--myself as related to others,--in other words, Person, Nature, Personality.

These are the elements of every human life. Their interplay is the story of any human soul.

There, for instance, is the story of every soul as told in a few immortal words,--the story of the lost and forgiven son. He said, "Father, give me what belongs to me;" there is one point of view, the person,--"what belongs to me;" and from that point of view he did whatever he wanted, "and wasted his substance in riotous living."

But "when he came to himself,"--there is a second point of view, his true nature, his true and higher self. Then he said, "I will arise and go unto my Father, and say unto him, 'Father, I have sinned against heaven and in thy sight;'" there is the third great point of view, himself as related to others.

That is the story of the human soul.

The indisputable facts dawn upon us, then.

Nature, person, personality, these three simple, essential elements of life in human beings, gather all the possible points of view or centres of personal consciousness in human life into three all-inclusive, constant centres of personal consciousness in one being.

Nature, person and personality are themselves therefore the three all-inclusive, constant centres of all personal consciousness in one being in man.

And these three centres of consciousness in man reflect, as our entire study has shown us, all the distinctions and relations of that vast, mysterious Divine Triunity of Scripture.

"The Whole of Personal Life"

In self-realization we saw that there are three distinctions, "I who see," "I who am seen," and "I who am known by myself." Those are, as far as they go, like the Three in the Three in One. But they are dim. And they have very few characteristics. And they are not very conscious. Do Nature, Person and Personality cover also those dim distinctions in self-realization, and make them real and vivid, and clearly conscious, and with characteristics like the Three in One?

Yes.

The Nature is the innermost, fixed and supreme vantage point from which I see all things. I see all things, unavoidably, from the view-point of my own nature. My nature is the primal "I" which sees. It is "I who see."

The Person is always that which is seen. It is that which is seen by others. It is that which is seen by myself. It is "I who am seen."

Personality is that by which a Person is known. It is I as I touch, affect and influence others and am known by them. It is equally that through which I know myself. It is I as I touch and affect and influence myself as well as others. It is the "I who am known by myself."

Those three centres of personal consciousness,--Nature, Person and Personality,--cover, as we saw a moment ago, all the points of view in inward decision, or self-direction,--and so make them always three,--and only three,--and marvelously like the Three in One.

And we can see that Nature, Person and Personality cover the three dim distinctions in self-realization,--and make them vivid and real,--and centres of consciousness,--and openly and clearly like the Three in One.

"A Perfect Human Likeness"

It hardly needs to be said that there is no thought of calling Nature, Person and Personality three persons in one human being. They are three centres of consciousness in one person. They are a purely human and limited likeness, in purely human terms, of the Three Persons in one Being in the Triunity of the Bible.

But that is exactly what a perfect human reflection should be. It is what an exact translation of such a Divine Triunity into terms of finite human life should be.

Three Persons in one Divine Being means three centres of Personal consciousness at once. It means Three who think at once, who know simultaneously, who act at one and the same time. It means Three who are conscious at one and the same time, and all the time.

The perfect reflection of that in man must be finite. Finite consciousness, such as man's, is limited by laws of succession and of one thing at a time. He thinks one thing after another, and from one point of view after another. He has three great centres of consciousness,--nature, person and personality. But as centres of consciousness they are always successive. He shifts from one to another of the three. These shiftings may be, and are, constant, rapid, and endless. But they are always successive. A perfect human reflection of Three Persons in One should reflect the distinctions and relations of those Three. They should include all the conscious points of view in human life. They should be constant and invariable, the very structure of human life. But as centres of consciousness they should be always successive, not three simultaneous persons in one being, but three invariable points in one consciousness. That perfect finite likeness of the Biblical Triunity of Father, Son and Holy Spirit is what human life marvelously presents.

As in the study of the physical universe and God, two great conclusions emerge again in the study of man and God.

1. The Triunity of Father, Son and Holy Spirit in God the Creator as described in Scripture presents truly an infinite and adequate reason for man's being what he is.

No other reason is needed. Above all other created things, man finds reason and cause for the pattern of his being in the Triune Being of God in whose image man exists. Man reflects everything else about God. His whole being seems to be a reflection of the Divine Being. Even sin cannot conceal this fact. Now we see that above all he reflects such Triunity as Father, Son and Holy Spirit. That explains man's pattern of existence. There is nothing else about God which man so marvelously reflects. There are men who are without love, to reflect God's love. There are men who seem to be without imagination, to reflect God's omnipresence. There are men who seem to be without will, to reflect God's omnipotence. There are men who seem to be without conscience, to reflect the holiness of God. But there are no men without nature, person and personality, however distorted or sinful those things may be, to reflect the Triunity of God. And there is no man who so absolutely reflects the love, the omnipresence, the omnipotence, the holiness of God as every man reflects the Triunity of God.

Perhaps this is why the human soul has always so instinctively and easily accepted the mysterious Triunity of Scripture, just as it has accepted God's love, or wisdom, or power, or goodness. It is because the soul is made in His likeness. It is not because we are thoughtless, but because we are profound. It is because we are like that Triunity ourselves. And perhaps that is why the simple, intuitive vision of children, and of savages, and of the untutored, and the profound intuitive vision of apostles, and saints, and great thinkers, and the simple and profound intuitive vision of every human heart which follows its instinct, has always so readily grasped the teaching of a Triune God. For the soul was made like Him that it might know Him.

2. On the other hand, man, in the absolute likeness of his essential being to the Being of God,-- Father, Son and Holy Spirit,--as described in Scripture, is the highest and most exact confirmation in the universe as we know it of that Divine Triunity.

For such Triunity of God, which ought to be reflected in His universe, ought above all to be reflected in man who is made in His image. And so it is. For man presents in his being a marvelous likeness of that Triunity of Father, Son and Holy Spirit. At every point, in every detail, the essential being of man parallels that Triunity. Even in three centres of personal consciousness the being of man rises into complete finite likeness of that infinite Triunity. It is supreme confirmation.

It is the method of science. The scientist says, "If such and such a great principle exists in the universe, it will be confirmed by its presence in such and such a place, and in such and such ways." The astronomer argued that the deflection of the planet Uranus from its exact orbit was due to the influence of an unknown planet. He figured that if this were so, then that unknown planet should be discovered exactly at a certain place in the solar system at a certain time. And it was discovered exactly there, confirming the principle. And so we say, "If such a Triunity as Father, Son and Holy Spirit, Three in One, exists in God, it will be confirmed, not only as we have seen it confirmed by the physical universe, but by a similar triunity in finite personal life in man, who as a personal being is made so much in God's likeness." Is it confirmed there? Remarkably. More precisely than anything else about God is confirmed in man's being. More exactly than anything else about the infinite is confirmed by the finite. The method of science, of confirmatory proof, is triumphant in this operation. Never did it work more successfully. An extraordinary Triunity was presented in the Bible and in Christian experience. It was corroborated elaborately by the physical universe. Now it is confirmed again, where it most ought to be, by the same principle exactly translated into finite terms in human life.

"The heavens declare the glory of God." Space, matter and time reflect the Three in One. "And God said, 'Let us make man in our own image.'" Even so He made him, the marvelous mirror and confirmation of the Triune God.

III

THE DEMAND OF THE UNIVERSE

The Challenge of the Facts--The first demand, and its astonishing answer--The Equation of the Universe--The second demand--"An invention?"--"A speculation?"--"A tendency?"--The third demand--The Axioms of the Being of God--The Vast Demand of the Universe--Not a philosophy of God but a philosophy of the universe--The Test of the Facts--The Challenge of Absolute Precision--The Converging Methods--The Three Sublime Conclusions.

ALL that we have been doing is, as we have said, the common method of science. The method is well-known. First, you correlate the facts.

Then, you seek an explanation of them. In that search you find a hypothesis which seems to fit and explain the facts. Then you carefully compare the facts with the hypothesis, to see whether they agree with and confirm that explanation.

This is what we have done.

We got together the elemental facts of the universe. We found a remarkable triunity in them. We sought the explanation. We found a similar Trinity attributed to the God of the Universe.

It was a mighty hypothesis. It was where it ought to be, in the God who is the Cause and Ground of the universe. It was found in the Bible, which tells so much else about God to which the human heart has assented. It was a hypothesis supported by the testimony of millions of the most thoughtful men and women of the race, who claim daily experience of Father, Son and Holy Spirit. It is the experience of Dante and Shakespeare, of Newton, Kelvin, and Pasteur, of Beethoven and Bach, of Michelangelo and Leonardo, of Charlemagne and Alfred, of Hampden,

Cromwell and Gladstone, of Washington and Lincoln. It is an experience in no way contradicted by those who have never had it but have never sought to have it.

With this mighty hypothesis seemingly fitting the facts, we sought corroboration. We did it by the most exact, minute comparison. We found perfect corroboration, flawless and complete. It was where it ought to be, in the structure of the universe. We found it at every possible point of that structure. It was truly a most extraordinary likeness of such a Triune Creator.

In Space, the universe showed an absolute likeness of unity, of threeness, and of three modes of being in that Divine Triunity. In Matter it showed yet more, extending into remarkable detail.

In Time, the universe showed a complete likeness of everything in that Triunity except three centres of consciousness. In human life the universe showed a remarkable likeness of every detail of such Triunity, including even those three Divine centres of consciousness.

We had said, "If God is Three in One,--Father, Son and Holy Spirit, as Jesus and the New Testament and the experience of Christians in all ages and lands present Him to us, it will be corroborated and reflected in His universe." We found that it is so corroborated, "beyond all that we could ask or think."

The Greater Challenge of the Facts to a Candid Mind

There is no real question as to the validity of this method of hypothesis and corroboration, when the hypothesis obviously first fits the facts, when it is supported by the personal experience of millions of honest and thoughtful witnesses, and when the hypothesis, finally, is universally and minutely corroborated by a comparison with all the facts. In science such a hypothesis so corroborated becomes a law.

Yet there is a still more vital way of handling the facts. That way is to begin again with the data, and to let the logic of universal facts carry us directly to their own conclusion, whatever it may be. That can be done, where the facts are universal and in complete unity, and where the mind which considers them is a candid one.

For this one must put aside one's preconceived ideas, or one's prejudices, or what one has been taught, or what one has been teaching, and gaze upon realities only.

May we do this now?

Here are the common facts of the world about us. They are so great, so universal, that we cannot wholly grasp them. But they are so simple, so broad, so self-evident, that we cannot for an instant reject them. Here are the facts which together make up the structure and substance of the universe.

Here first are the facts of space. They are self-evident. Height, length and breadth, with all their characteristics and relations, as we have sketched them, are evident to all who will think about them. No one can dream of questioning them. They are so evident, that they are the basis of the exact science of geometry.

Here also are the facts of matter as we have seen them. Here is the whole structure of matter, in energy, motion and phenomena. They and all their characteristics and relations, as we have recounted them, are self-evident. No one will question them. They are the basis of all modern science.

Here further are the facts of time as we have seen them. Here is the whole being of time, in future, present and past. Those three elements in time, with all their distinctions and relations, are self-evident. No one will question them. They are basic in all human thinking and human experience.

Now we go further.

Through all these universal facts which compose the physical universe there runs a remarkable triunity. They are each of them a remarkable triunity. They are the same kind of triunity. All of them together constitute a universal triunity, a tremendous and unescapable fact, the whole structure of the physical universe.

Now we look from the physical universe into human existence. We find a remarkable thing. We find that exactly such a triunity is here, also. We all know certain broad and simple facts which include the entire structure of human life. No one can question them, or question the self-evident distinctions and relations expressed in the simple, universal words "nature," "person" and "personality." Human life is exactly such a triunity as the physical universe is. Whether one values them or not, whether one sees in them a higher meaning or not, these are the absolute facts.

We face, therefore, a further great fact, a yet more universal one. No one can question the self-evident likeness to each other which runs through all these things which are the structure and substance of the universe of space, matter, time and human existence. They need only to be put side by side. A strange, universal triunity of height, length and breadth, of energy, motion and phenomena, of future, present and past, of nature, person and personality, is visible. It is the very being of all these things. It is so exactly alike in all these forms, so complete, so flawless, that it is identity in terms of space, matter, time and human existence. This universal, common triunity, once pointed out, is evident to anyone who knows anything of space, or matter, or time, or human life.

These are the facts. What shall we say of this structure of the universe? This general triunity in all things requires an explanation. It demands a reason. So great and universal a phenomenon cannot be dismissed by any student of the physical world or of human life. There must be an explanation, if this is an orderly universe.

And that it is an orderly universe is the first principle of all science.

There must then be a cause and explanation of this universal triunity. It must be a great and universal cause. Nothing less is possible. Something in the common cause of space, matter, time and men must be the common cause of this triunity.

There is but one common cause of space, matter, time and men. Whatever methods one may uphold, there is but one common cause. That cause is God. He is the Cause of the universe.

This universal triunity in space, matter, time and men demands an explanation, then, in the Cause of the universe. It demands a cause in God. And by this we mean, of course, not an arbitrary or accidental cause, but a deep and basic cause, in God.

With this demand of the universal triunity presenting itself from all directions, we come to another Fact. Jesus of Nazareth and the New Testament present exactly such a cause in God as the universal triunity demands.

For the Triunity of Father, Son and Holy Spirit there presented is an absolute counterpart, in terms of God, of the universal triunity in terms of space, matter, time and man. It presents exactly such a cause as the universe demands.

It is a cause in God.

It is not an arbitrary or accidental cause, but a cause deep in the whole being of God.

It is in God who is the common Cause of space, matter, time and man.

It is a vital and adequate reason for His making the universe just what it is.

It is an exact original for such a universe of tri-unities to reflect.

It is a complete response to the demand of the universe.

In every detail and in every broad outline this Triunity brought by Jesus and the New Testament satisfies every outline and detail of the triunity in space, matter, time and man. This is true even of the most mysterious details of the New Testament Triunity, and of the purely physical things of the universe. Then this is what we face:--

The universal triunity demands at the heart of the universe a certain Triunity as its Cause and Original in God; and the Bible brings us exactly such a Triunity in God.

It is an appeal to our candor. We must put aside prejudice, or any fear of what is new and vast. There before us are the facts of the universe. Here is the description in the Bible. We cannot, if we would, evade the question.

How did this Biblical Triunity come there,--so exactly, minutely, immeasurably what the universe requires as its Cause,--if not from God, and from that Triune Cause which must be in Him?

Where else could the writers get it?

1. From nothing:--by invention?

Did they get it out of nothing, by pure invention, by a happy coincidence which made their invention correspond with the vast triunity in the physical universe and in the universe of the soul?

That is a suggestion which will not be considered long by any well-balanced mind. A coincidence embracing every distinction and relation in the entire structure of space, and of energy, motion and phenomena, and of future, present and past, and of nature, person and personality, and every distinction and relation, the most detailed and mysterious, described in the Divine Triunity of the New Testament,--a coincidence without a slip or a flaw in all this vast and complicated fabric,--no mind accustomed to the correlation of facts will really consider for a moment a suggestion so naive or so fantastic. We can dismiss this at once. Such things as these are not coincidences, in an orderly or sane universe.

The New Testament cannot have gotten this Triunity out of nothing by pure invention, presenting by mere coincidence exactly what the universe of space, matter, time and men requires as its cause.

2. From Man,--by human instinct?

Did the New Testament writers get that Triunity of God from within themselves? Was there some instinct in them, which led to such an idea of God?

Some say that there is a tendency of the human race to see God as triune, that other religions show it, and that the New Testament writers got their idea of God from the same tendency.

If there were such a tendency, it would be another and profoundly convincing groping of the human soul after the Triunity in whose image it is made. It would be an extraordinary evidence of the reflection of that Triunity deep in the being of the human soul.

But is there such a tendency? Do other religions show such triunities? If they do, are those triunities sufficiently like the Biblical Triunity to show that this Triunity came from the same tendency? And if there are such triunities in other religions, do they, like the Biblical Triunity, meet and satisfy by their exact resemblance the demand of the vast triune fabric of space, matter, time and human existence for a triune Cause?

There is only one answer. We have not found such triunities in other religions of the human race. One finds triads here and there in a few religions. But they are not essential triunities. They apparently happen to be groups of three. In some religions there happen to be larger groups, such as Odin, Freya, Thor, Loki and Baldur in the Norse religion; in some religions groups of two, as Ormazd and Ahriman in the Persian religion. Such mere groupings of three which are not organically one do not, of course, make a trinity, or show such a tendency in human thought, any more than other human groupings, such as three wheels in a tricycle, or three singers in a trio, or three officers in a boat, or three days of grace in business, or three bases in a game of ball. Such efforts to find a likeness of the Trinity of the Bible do not get one anywhere, whether they are the efforts of believers in that Trinity to find proofs of it in anything which happens to be three, or whether they are the equally naive efforts of unbelievers in that Trinity to find disproofs of it in anything in other religions which happens to be three. Such efforts in comparative religions, such explorations in moonlight and cobwebs, land the explorer in nothing more intelligent than a professor's chair. For a sympathetic study of the gropings and yearnings of the human soul through the best in human religions discloses nothing which shows any tendency toward genuine triunity in the idea of God. They show no resemblance to the Triunity of the Bible which would indicate a common origin. They have nothing which brings an answer to the demand of the universal triunity for a cause in God. The grotesque three in Egyptian mythology offer nothing except to fond credulity. Do the three in Hinduism reveal a trace of triunity, and of resemblance to the Triunity of the Bible?

If they do, it is very faint. And if they do, they do not necessarily show an original tendency. For the triad of Brahma, Vishnu and Siva was a later development in Hinduism, a long time after Christianity, with its Triunity, had been active in India. The older Hinduism showed no trace of triune tendency in religion. But the triad of Brahma, Vishnu and Siva has no real triunity, no real likeness to the New Testament Triunity, and no meeting at all of the demands of the universal triunity for a cause in God. Truly there are no evidences of a natural tendency of the human race to create such triunity in God. If there were such a tendency, enough to account at all for the Biblical Triunity, it would show itself vividly in all religions. But no such tendency emerges. The human soul does readily respond to the presentation of the Three in One of the New Testament. There is an instinct there. But the religions of the human race reveal no tendency to depict God as triune. The writers of the New Testament did not get their picture of the Triunity of Father, Son and Holy Spirit from a general tendency within their own souls. Wherever they got it, they did not get it there. It did not come from them.

Did the New Testament writers get that Triunity from the universe, by speculation from the universal triunity? Would not this explain the likeness? Did the writers of the New Testament get their idea of a Triune God from the triunity in worlds and men, and translate it into terms of Deity, in Father, Son and Holy Spirit?

There is no evidence for such an origin. There is no evidence in the New Testament that the writers got the Trinity from the triunities in space, matter, time and men. But there would certainly be some such evidence if they had gotten that Triunity in that way. There are none of the indications which there would certainly be if such a hypothesis were so. None of the Biblical writings reveal any trace at all of such a thought or speculation in the mind of the writers. There is no attempt in any way to compare the Triunity of Father, Son and Holy Spirit with space, or matter, or time, or anything in man. The writers never speak of time as future, present and past, and never mention the distinctions in time in connection with the Trinity. They never mention, and there is no reason to think that they knew, the triune nature of matter in energy, motion and phenomena, the discovery of modern science. There is no hint in the New Testament at any such triunity in man. The triad of nature, person and personality is never suggested. There is no mention of a relation of the Divine Trinity to any of these things. Indeed there is no theoretical Triunity in the New Testament. The presentation of the Trinity is simple, natural, matter-of-course, a phrase here, an allusion there, now a characteristic, now a relationship, as it happens in connection with other things and topics. Indeed the Trinity there occurs largely in the sayings of Jesus, and in his simplest, most personal talk about his Father and himself, and about the Spirit.

There is no evidence at all, then, in the New Testament presentation of the Trinity of any thought about triunities in the physical world or in man, nor of anything theoretical at all. There is no evidence, such as there surely should be, to indicate that it came from the triunities in space, matter, time and man. This Trinity of the Bible evidently did not come by human speculation to be so exactly what the universe requires. No one can think that it did. It is clearly not a theoretical invention of the New Testament writers, but is an honest presentation.

Only One Source is Left

The New Testament writers did not get that Triunity from nothing, by pure invention. They did not get it from man, by instinct. They did not get it from the universe, by speculation. Only one source is left.

It can have come only from God, and from that very Triunity which the universal triunity reflects and demands as its Divine Cause.

Absolutely Reasonable

When we compare the universal triunity at its height in personal being in man with that Triunity of the Father, Son and Holy Spirit which it demands, we see how absolutely reasonable all of this is. It grows with analysis.

For Nature, Person and Personality, as we have seen in an earlier chapter, are just what a finite likeness of the Triunity of Father, Son and Holy Spirit should be. They are three all-inclusive centres of personal consciousness. They are not simultaneous, but finite, and always successive. Man thinks and feels one thing after another, from one point of view after another. This is because he is finite and limited. But this is what a finite

reflection of three simultaneous Persons in one Divine Being should be. And these three constantly successive personal points of view or of consciousness reflect completely in a finite way all the distinctions and relations found in the Divine Triunity as described in the New Testament.

On the other hand, the Triune Father, Son and Holy Spirit in God are just what an infinite Original of human Nature, Person and Personality should be.

For a Divine equivalent of Nature, Person and Personality would be, as in their human reflection, the all-inclusive centres of all His Personal consciousness.

But in Him they would be, not successive as in us, but simultaneous. God is bound by no laws of one thing at a time. He thinks many thoughts at once. He knows all things at once. He is all-knowing and all-conscious. His omniscience means that He should not be bound and limited as we are, to one thought at a time, to one point of view or centre of His consciousness at a time. He should be conscious simultaneously in those three centres. They should in Him be three great centres of Personal consciousness, at one and the same time, and all the time, in one Being. But three simultaneous, constant, Personal centres of consciousness in one Being is the very definition of the Trinity.

And if these three Personal centres of consciousness in God have the relations and characteristics which nature, person and personality have in us, they will be exactly such as Father, Son and Holy Spirit as described in the Bible.

Those three Persons in one God are just what a Divine Original of the human reflection in Nature, Person and Personality should be.

The Riddle of the Universe Becomes the Challenge of the Universe

Very clearly then the challenge of the universe, the challenge of the vast ascending triunity of space, matter, time and human existence, faces and surrounds us.

The entire triune universe demands that its Cause and Original should be a similar Triunity in God.

The New Testament brings us exactly such a Triunity.

The universal triunity demands that very Triunity of the New Testament, which can have come only from that Divine Original which the universe reflects.

The universal triunity, rising to its height in personal being in man, declares that that Triunity of Father, Son and Holy Spirit is exactly what the Original of the triune reflection in man should be.

This is the result when we let the facts of the universe tell their own story and carry us where they will.

You can see this as overwhelming, demanding, universal, exact confirmation of the Triunity of God in the Bible.

Or you can see it as the challenge of the facts of the universe to any open mind.

A candid mind must face this confirmation and challenge. No prejudice, or old habit of thought, or preconceived ideas, can stand in our way, when God reveals Himself.

And when once in candor one has accepted the revelation and the Fact, one must let it ever afterward mould and color all one's thinking. Either all this revelation is true, or it is not true. If it is true, it must illumine all one's universe. One cannot put it in a corner of one's mind, with other secondary facts. If it is true it is supreme. It must affect all one's science, one's philosophy, one's esthetics, one's outlook upon the world and life, and one's personal religious experience.

For surely no mind whose doors are open to the voices of the universe will attempt or desire to disregard their deep but overwhelming chorus. No mind blest with the gift of vision can be wholly unmoved by so marvelous a universal mirror reflecting, confirming and demanding that Triune God.

Not a Philosophy of God, but a Philosophy of the Universe

Again we should remind ourselves of one thing.

Nothing which we have been seeing or saying shows that God had to be Father, Son and Holy Spirit, or why He is Father, Son and Holy Spirit.

No argument that God must by a necessity of his nature be so and so because man is so and so carries any real weight. For as we have said, "God is not what He is because of anything that man is."

What we have seen is not an argument for the reason and necessity of God's Triunity. It is a revelation of the Divine Reason and Cause of man's triunity. It is the testimony of the universal reflection to the Divine Original.

When we look at the universe and say "There must be a God", we do not mean that God had to exist because of the universe. We mean that we must recognize that there is a God, because we see the universe.

So when we look at the universal triunity and say "God must be Triune," we do not mean that He is or

had to be Triune because the universe and man are triune. We mean that we must recognize that He is Triune, as the New Testament reveals, because we see that His universe, which reflects Him, and man, who reflects Him, are both triune in the likeness of that Triunity.

You do not exist, nor appear as you do, because of your image in the mirror. But your image in the mirror is overwhelming evidence of your existence and of your appearance.

But it is not an "argument" for an inherent necessity in God's nature. We might well give up all such arguments from something in man's nature to an inherent necessity in God's nature. God is not what He is because of what man is. God cannot be explained in that way. There is no explanation of the Trinity.

So if someone asks, "May we at least find in nature, person and personality an explanation of the Trinity? May we not have in them at least a philosophy of the Trinity?" the answer is brief but complete. We do not find, in all these things, any basis or philosophy of the Trinity, because the Trinity is not based on them; they are based on the Trinity. The human nature, person and personality cannot be seen as in any way the basis of Father, Son and Holy Spirit, for the simple reason that man's nature is not the basis and cause of God's nature. The order is the other way. Man is not the cause of God. God is the cause of man. The created nature is not the cause of the Creator's nature. The Creator's nature is the cause of the created nature. Man is not the original, and God made in man's image. God is the original, and man is made in God's image. We argued not from original to reflection, but from

reflection to original. Father, Son and Holy Spirit are not what nature, person and personality are in God. They are what is in God in place of nature, person and personality as we know them in man. The Trinity is not the reflection, spread large and infinite in God, of nature, person and personality. Nature, person and personality are the reflection, made human and finite, of Father, Son and Holy Spirit.

Nature, person and personality are not therefore an explanation or philosophy of the Trinity. There is no philosophy of the Trinity. The Trinity is the explanation and philosophy of the triunity in man, as it is of the whole triunity in the physical and human universe. We have not, therefore, in these things, a philosophy of the Trinity. It is an explanation and philosophy of the universe. It does not tell us why God is Three in One. We do not know why God is Three in One. It tells us why the universe is triune. No man has a right to make a philosophy of the Trinity out of it. It is a philosophy of the triunity of space, matter, time and man, which is the fabric of the universe. God is the explanation of the universe. The Trinity explains and alone explains the universe of such triunities. It tells why space is what it is. It tells why matter is what it is. It tells why time is what it is. It tells why man is what he is. It is the explanation, and the only explanation, of the universe of triune space, matter, time and man.

The Test of the Facts

How simply and naturally all this comes to us in the New Testament! How unphilosophically and untheologically! It is nowhere presented as a doctrine. It is everywhere present as a fact. It comes to us above all in Jesus, in what he says and does. It shows itself in all that he tells of himself, and of the Father, and of the Spirit.

This simple and natural realism is shown to us by one great test:--

The Trinity is absolutely involved in Jesus. It is never presented apart from him, either by himself, or by the New Testament writers. It is his claims, as the Son, the second Person in that Three in One, and his allusions, and the claims of the New Testament writers about him, which have brought the Trinity to us. There is no Trinity apart from Jesus. There is no revelation of the Trinity except through him.

But he is a man. He is the most real and human of men. In his interests, his sympathies, his social life, his friendships, his joys and sorrows, his emotions, his life and death, he is the most human of all men, the most natural, the most unartificial, the most free from the poses and pretences and aloofness of great men.

Now we have this situation. As a human being he should have, and we should find in him, nature, person and personality of the most human kind. As the Son, the Second Person in the Three in One, we should find in him and through him Father, Son and Holy Spirit. We should find a remarkable thing. We should find the two, the Divine Trinity and the human triunity, coinciding and blending in him. The Divine Father, Son and Holy Spirit and the human nature, person and personality should, in him, be seen merging and uniting. This is what we should see.

Is this what we actually do see? Is it what we see and hear in Jesus as his life and words are brought to us in the New Testament?

First of all we know Jesus as a person. We find him so in the Gospels. He was a very real person in every way. There was nothing abstract about him. To everyone who knew him or met him he was very genuinely a person. There never was any question about that.

Then as his friends got to know him better they saw back of the person his very nature. They found it to be a Divine nature. It showed itself so to them in a hundred ways. Then it came to them by his words and by their observation of him at work and at prayer that back of and in that Divine nature was "the Father." "The Father abiding in me doeth the works." "This is my beloved Son." "I and the Father." "All things have been delivered unto me of my Father, and no one knoweth the Son, save the Father; neither doth anyone know the Father, save the Son, and he to whom the Son wills to reveal him." Jesus "dwelt among us, and we beheld his glory, glory as of the only-begotten of the Father."

And all felt the influence of Jesus' marvelous personality. It entered other lives. It gave them new life. It made them over new, so much so that they were said to be born anew. It was vivid and real beyond all experience of personalities. Finally he taught them that this life of his as it entered, and influenced, and transformed other lives, this personality, this spirit of his, was more than a personality. It was also a Person, his other Self, the Holy Spirit. And so it developed that when he, the visible Person Jesus, was gone, this Personality of his proved to be a distinct, invisible Person, the Holy Spirit, who spoke, and directed, and acted, and said "I" and "Me."

This is exactly the representation of Scripture, in its day-by-day narrative of Jesus and his friends. In Jesus we see, before our eyes, nature, person and personality reaching up into Father, Son and Holy Spirit and merging with them. Indeed it was this very blending which first involved the Trinity. It was the presence of the Father in Jesus, where we would have simply our own nature! It was Jesus sending out the Spirit, where we would send out simply our own personality! It was these very facts which first brought the Trinity to the world and to us. It was in this simple, this human as well as Divine, this natural way that the Trinity of Father, Son and Holy Spirit came to the knowledge of men.

Simply, self-evidently, naturally, without explaining or theorizing, it all comes to us, and fits together, and lives before us, in the life and words of Jesus in the New Testament. That seemingly impossible blending of the Divine Trinity and the human triunity becomes in Jesus' daily life and conversation a simple reality. The insurmountable problem becomes a living proof. The data in the case fit perfectly the great conclusions and evidences of the universe about the Triune God. The facts of Jesus meet and fit those mighty facts of the universe which require and demonstrate the Father, Son and Holy Spirit.

The Challenge of Absolute Precision

Indeed we find such fitting-together at every point, in the most scientific way, of all the facts in the case. This absolute exactness is an overwhelming challenge, a demand, confronting every open mind. If God is Three in One, Father, Son and Holy Spirit, there ought to be such and such triunity in the physical world. And there is such triunity, in the whole structure of the physical world, with such absolute and elaborate likeness to the Divine Triunity as we could never have asked for. If God is Three in One there ought also to be such and such triunity in human existence. And there is, as the very being, simple and self-evident, of human life, with such endless and absolute detail of likeness as no one could have dreamed of asking. Or on the other hand, the cause of all this triunity of the physical world and man ought to be found in God. And exactly such a cause, complete, exact, profound, is presented to us in Father, Son and Holy Spirit in the New Testament. And further, if human existence is Nature, Person and Personality, the Original of this reflection should naturally be three simultaneous, conscious Persons with such and such characteristics. And it is exactly as the Bible describes Him. And finally, if God is Three in One, and Jesus is therefore God and man, then that Triunity in God and this triunity in man ought in Jesus, as God and man at once, to be seen merging into each other. And that is exactly what we do see, and what first drew the world's attention and ours to the whole question of the Trinity.

The Converging Methods

Whether we follow one method or the other makes no difference. The result is the same. The two methods converge. We can begin with the Trinity, presented in the Bible, and experienced by countless men and women, and summon the facts of the universe to reflect and confirm it. Or we can begin with the facts of the universe, and see them demanding the Trinity as their original and cause. One can look upon the visible world. Or one can look into the secret, invisible forces of nature. Or one can look within the inner universe of one's own being. Wherever one looks one finds that marvelous Triunity.

Whither shall we flee from that Triune presence? If we ascend up into heaven, it is there. If we take the wings of the morning, and dwell in the uttermost parts of the universe, it is there. If we say, "Surely the darkness, and the invisible things of the universe, shall cover us;" even the darkness hideth not from that Three in One. The heavens declare the glory of the Triune God, and the firmament showeth His handiwork. For the invisible things of Him are clearly seen, being perceived through the things that are made. And God said, "Let us make man in our image, after our likeness." And God created man in His own image. In the triune image of God created He him. Without, within, above, below, whichever way we turn we see the Triune God.

The Three Sublime Conclusions

Out of this candid study of God and the world and man rise three conclusions so great that we could not grasp them if they were not also so self-evident.

1. The universe is one vast evidence of that Triunity of Father, Son and Holy Spirit in God. The entire universe, the outer universe of space and matter and time, and the inner universe of the human soul, in all itsvast triunity, reflects that Triunity. It demands that Triunity in God.

The universe is one vast evidence in such detailed, exact, scientific correspondence to the Divine Triunity which it demands and reflects as can be found in no other witness of the universe to anything else about God.

And if you are one who feels that space and time are simply ways in which you have to see the universe, thought-forms in which you must conceive the universe, then surely you, who must always see the universe in such triunity, cannot hesitate to see God as such Triunity.

2. The Triunity of Father-Son and Holy Spirit is the explanation of the universe. It is the answer to those great questions, "What is the explanation of the universe?" "What is the principle of the universe?" The answer is "Triunity in the image of the Triune God is the principle and explanation of the universe. It is the organizing principle of all things. It is the structure and pattern of the universe."

Why is space as it is, sheer, inevitable, intangible threefoldness?

Why is this a space universe?

Why are energy, motion and phenomena the being of matter?

Why is the universe as it is in this way?

Why is all existence a continuity, a continual passing out of future, through present, into past? Why is this a time universe?

And if space and time are forms of thought in which we conceive the universe, why do we always conceive it so? Why are our minds such that they conceive it so?

Why is man exactly what he is? What is his likeness to the universe?

What are the reasons for this?

Is there one great reason for all? Are all the reasons one? Is there one great explanation? Is the formula of the universe beyond us? Does it need a "higher mind" to formulate it? Or can we grasp it?

The answer to all these questions is evident. The formula, the pattern, the secret, the principle, the structure, the explanation, of the universe is self-evidently clear. It explains itself. No "higher mind" is needed to formulate it. It clearly includes God. It is a universal principle. It is a principle which lies in God's nature. It is triunity in the image of a Triune God. It reveals one vast unity,-- sheer space, and moving matter, and mysterious time, and wondrous man, and supreme God, bound in one vast unity.

Get in your heart a vision of the universe. Do not go about in the midst of it blind to the great structure of the universe and of human life. See its immense triunity. See it reflecting from leaf and star and space and time and self the Triune God. See it as an orderly universe. See it as a whole. See it as one vast, open and visible witness to the Three in One. Have a philosophy of the universe. Open your soul to the visible explanation of all things. Get a vision of yourself as a part of the universe, in a body of space and substance, reflecting in these the triune Creator,--living a life of future, present and past, in vivid triune likeness to Him,--existing in nature, person and personality, an image of the Triune God. Man was made to reflect Father, Son and Holy Spirit. Man was made to know and to commune with Father, Son and Holy Spirit.

And get one further vision, the focus of all of this logic and all these realities.

3. The universe is one vast witness to the claims of Jesus. That Triunity of Father, Son and Holy Spirit which it the whole universe corroborates comes to us in Jesus. It is bound up with him. That Triunity of Father, Son and Holy Spirit which supplies the universe with its cause and explanation comes to us in Jesus. It is revealed to us in him alone, and he is the visible embodiment of it.

As beings who live in this universe, and are a part of it, we cannot ignore Jesus, to whose claims and whose Sonship this whole universe of height, length and breadth, of energy, motion and phenomena, of future, present and past, of nature, person and personality, is one vast converging witness. Men crucified him because he made himself God. But millions experience the reality of his life and power to-day. No man with vision can ignore Jesus the Son. For we live in a world which is his vivid likeness, amid a universe of interwoven movement which is his seamless robe, and in a present which is his living reflection in the stream of time.

PART II.

Chapter 1

THE PROBLEMS OF THE UNIVERSE

INTRODUCTION. THE SECRET OF ALL THINGS

THE Triunity of Father, Son and Holy Spirit reflected and expressed in the triunities of space, matter, time and man is the secret of the universe. It is the key to the many riddles of the universe. It shows why things are as they are, and that they did not happen so. It shows why space is what it is, of three dimensions. It shows why matter is what it is, of energy, motion and phenomena, with all their relationships. It shows why time is as it is, composed of future, present and past. It shows why man is made as he is. It may show the great principle of unity in all things. It should illumine the relationship of space, matter and time. It may well cast light on the mysterious principles of existence, of change, and of reality, in the universe. The being of God, as the central fact of the universe, may explain these universal things. It may make them clear, not as we ourselves paradoxically try to make mysteries clear, by involved effort, and intricacies of thought, and abstruse analysis, but by the broad, self-evident Fact of a Triune God.

It is a world principle. It means for you, if you will, a new system of thought, a new way of thinking, based on the being of God and of His world.

It is, if you will, a true philosophy of things physical and things human.

But better, it is a new vision of the universe, in the light of the Triune Being of God.

I. THE SECRET OF THE UNIVERSE AND THE PROBLEM OF THE UNITY OF ALL THINGS

No mind can think long and deeply upon the universe without asking a very great question. What is it which all things have in common? What makes this a universe? What is the basis of unity?

We feel that there must be such a basis. Science and philosophy both seek such a unity in all things.

That basis is not the atom, nor the electron. Some have tried to find such a common basis of unity in the unit of matter. They have not only tried to explain the physical universe in that way. They have made even the human soul a combination or activity of atoms or electrons, or a series of reactions of mechanical forces. That effort is futile. The common basis of unity in all things cannot be found in such materialism. The attempt to put the soul on a physical basis will always seem and always be an artificial effort at unity.

Neither can that common basis of unity be seen as spirit. There are those who would see it so. They would view even the things of the physical universe as the apparently physical manifestations of pure spirit. Many philosophies and some religions have attempted it. But it cannot be done. The common basis of unity cannot be found in such unreal idealism. Matter cannot be explained as spirit. It is an artificial unity which such an effort brings.

What is the basis of unity? The problem is profound. The basis of unity cannot be a common stuff, or a common substance, either physical or spiritual. It cannot be the materialist's unity or the idealist's unity, either atom or spirit. What is it? The answer to this question must needs be a very self-evident one, and it must be something other than a common substance.

There is for us a clear answer and a universal one.

Triunity.

Triunity is what stars, and trees, and rocks, and water, and gases, and light, and heat, and space, and time, and the human body, and the human soul, and human daily life, and human consciousness, and human will, all have in common. Triunity, not vague and general, but of a very definite and exact kind, is what they all have and what they all are. It is a universal and unvarying basis of unity.

This triunity in the likeness of God, this basis of unity, is seen to be not a common universal substance, but a common universal structure. The unity of all things, the unity of physical and spiritual things, cannot possibly, as we have seen, lie in a common substance. This basis of unity in the universal triunity is not a common substance. It is a common structure. The search for a common, universal substance is a hopeless one. It can succeed only by making the material world spiritual or the soul physical. Triunity is far from that fallacy. It is not a common substance. It is a common principle and structure of all things.

Triunity in the image of the Three in One is that which all things have in common. It is the structure of all things. It is a basis of unity in all things. May we say, It is the basis of unity in all things. It makes this a universe.

II. THE SECRET OF THE UNIVERSE, AND THE PROBLEM OF SPACE AND MATTER

Is space the source of matter?--But what is space?--But what are dimensions?--"Real space"-- How does power come to be a physical universe?--The Reality of Space, Energy and Motion--

How modern discovery leads to this reality--How "ether" leads to it--The troubles of "ether''--The way out--Discontinuity--Continuity--Momentum--Magnetism--Vibrations--Quanta--Electricity--Electromagnetism--The universal energy--The true Continuum--Particles versus Waves--Space and Motion.

What is the relation between space, matter and time?

That question must have hovered in our minds as we have thought about the three. It is indeed a question which is very much around us in the air to-day. The question of the relationship, or, as people like to call it, the relativity of space, matter and time, besets any thinking man or woman in recent years.

What is the organic relationship of space, matter and time to each other? They are together the structure of the physical universe. That of course we know. They share a common vast Triunity, each of them being a marvelous embodiment of it. That also we know. But is there any other relation between them? Are these three triunities, which are the fabric of the physical universe, a yet vaster triunity with each other? Does then the Divine Triunity explain and illumine the relationships of space, matter and time to each other? There are surely many mighty things which depend on it.

Is Space the Source of Matter?

In the Divine Triunity the Father is the Source of the Son, and the Son is the embodiment of the Father. In the triunity which we call matter, energy is the source of motion, and motion is the embodiment of energy. In the triunity of time, the future is the source of the present, and the present is perpetually becoming the embodiment of the future.

In the triunity of man, the nature is the source, and the person is the embodiment. Is it so then with space, matter and time together? Is space the source of matter? And is matter the embodiment of space? Are the relations between space and matter those of the universal triunity?

We know what matter is. It is energy, motion and phenomena. Its chief factor is motion. But what is space?

But What Is Space?

What is space? The answer is at once simple and complex. The simple answer is self-evident. We can all agree upon it. Space is composed of dimensions. That is certain. Space is nothing else. It consists of dimensions.

That is the clear and obvious answer. But it is not answer enough for genuine reality such as we now would reach. We must go further.

What then are the dimensions which we call space? What are they dimensions of?

They are of course the dimensions of space. But that is tautological. It moves in a circle. They are the dimensions of space. But space, on the other hand, is composed of these dimensions. A certain building is built of boards. You ask, "Boards of what?" The answer comes, "They are boards of the building." Of course they are. But the building is composed of the boards. What are the boards composed of? Of oak? Of pine? Of maple? So space is composed of dimensions. But what are the dimensions composed of? What are they dimensions of?

Are they dimensions of nothingness, of absolute emptiness? That is a strange idea. But stranger still, it is a very prevalent conception of space! Space is often described as nothingness, and its dimensions as dimensions of nothingness.

But can nothingness have dimensions? You may indeed think of an empty box or chamber, and think of its emptiness as having dimensions. It might be emptied of air, and contain only a vacuum. Then there is nothing at all there, yet there are dimensions there. But in that case it is not the emptiness or vacuum which has the dimensions. It is the box or chamber which has the dimensions. How can sheer, absolute nothingness have dimensions or anything else?

The unreality of the common idea of space as nothingness in several dimensions is strikingly shown by the revolt against it. That revolt, as we know, regards space as being only a form of thought in which we conceive matter or motion. Kant of course is sponsor for the idea that space is a form of thought in which the mind conceives matter. But if in Kant's day it had been clearly known through the advance of science that matter is essentially motion, and that this motion moves through incredible distances with inconceivable rapidity, that great thinker would probably not have held that space is purely a form of thought. For motion would exist if minds did not exist to think about

it. And where motion is, there must be genuine extension and dimension. Motion needs room. Motion would create dimensions if there were none. Space is more than our way of conceiving motion. Unless all motion is a figment of the mind, the dimensions which we call "space" must be an outer reality.

But What Are Dimensions?

What then are these dimensions? What is the reality of space? Are they the dimensions or measurements of motion? Is that the ultimate reality, the basic fact, of space?

No. For motion is not basic. Back of motion is energy, which passes into motion.

Is energy basic, then? Is that the primary thing, so that the dimensions of space were and are the dimensions of energy?

No. For back of energy in the universe is whatever produces energy. Wherever energy is, there must be before it the power which produces that energy. There must be creative, causal power outspread. Before energy and motion were spread out in the universe, there was, there had to be, the outspread, omnipresent power which produces energy and motion.

The dimensions which we call space were not primarily then the dimensions of energy. They go further back than that. They were the dimensions of that which produces energy and motion. They were the dimensions of outspread, universal, omnipresent Creative power, in a universe with God in it.

Is that ultimate? Did the dimensions which we call space exist before there was the presence of that Creative power? Was there ever a time when Divine power was not outspread and omnipresent, so that the dimensions of space were first of all the dimensions of nothingness? Are we thrown back after all upon that contradictory idea?

There is no need of that impossible conception. For there never has been a time when there was nothingness and not God in the universe. There was never a time when His omnipresent power was not outspread everywhere. There is, therefore, no need to imagine that absurdity, a vast outstretching of nothingness in three dimensions. For there has always been the outspread power of God. The attempt to imagine the dimensions of space as the dimensions of nothingness is really an attempt to imagine a time before God existed, and to picture a vast emptiness in which His power was not yet present. But with an eternal God, and a God always omnipresent, and therefore with His power always present everywhere, the dimensions of the universe, which now are visibly the dimensions of motion, have always been primarily the dimensions of that Divine outspread power which precedes all energy and motion. Those are the true and basic dimensions which compose space.

"Real Space"

These dimensions are a reality. They are real space. Space is not the absolute unreality which some would picture,--the outspreading of primal nothingness. Nor is space that practical unreality,--the outspreading of the mind to perceive matter or motion. For motion must precede its perception by the mind. And energy precedes motion. And Creative power outspread in dimensions precedes energy.

Space then has always been the extension, not of the human mind, nor of motion, nor of nothingness, but of God's power.

Space is reality. It is not nothingness in three dimensions. It is not a figment of the mind as it gazes upon motion. It is reality.

This is the reality which we get from the principle of Triunity as it points to space as the source of motion.

How Does Power Come to Be a Physical Universe?

But how does the outspread Creative power whose dimensions are space pass into a world of matter or motion? How do we get a physical universe? God's power is surely first of all spiritual. God is a spirit. How does Divine power become physical? How does it pass into a tangible universe? That is one of the great questions of thought.

But the answer is clear. God's power is not only power to think but power to move. Can spirit move in a spatial way? Surely. Your mind is spirit, and it can leap across the sea, and pass to planets and stars. So, surely, Creative power can move. But, as power to move, that outspread Divine power becomes energy, which is the power of physical motion. And this energy not only can move, but, as we know, does move. It becomes motion, everywhere, which manifests itself in all the phenomena of matter. It becomes a tangible universe, which we can feel, and see, and hear.

The Reality of Space, Energy and Motion

Space then is, not in some vague way, but logically and truly, the basis and beginning of the actual, tangible universe. It is not a framework, a location, of vast nothingness, in which in some way the universe is built. It is itself truly the basis and beginning of the physical universe. For this real and living space, which is the outspreading of Divine power into dimensions, is that by which such Divine power translates itself from spirit into energy and motion, and so into a physical universe of energy and motion.

But people may hesitate. This is not the way in which one has always thought about space, if one has been able to think about or conceive space at all, either as physical nothingness or as mental unreality.

This reality of space is however far more reasonable than the contrary and self-contradictory idea that space is nothingness, or a figment of the mind. It gives living reality to space and to the physical universe.

And there is no reasonable objection to it.

Can Divine power, we may ask, the power of God who is a Spirit, have dimensions? Surely. Omnipresence in a physical world must have dimensions.

To some, too, this vision of space as the outspreading of Divine power may seem a more religious view than they are willing to sanction. But such a view of space is hardly unfitting in a theistic universe, a universe with God in it, the only sane universe to a sane mind.

And it may be well to bring to mind how surprisingly this view of space fits in with the modern view of the physical universe as a universe of infinite activity, whose realities are not things or forms, but energy and motion. For nothing is inactive in the universe as we now know it--not even space!

How Modern Discovery Leads to This Reality

For this conception of space as the outspreading of Creative power, which passes, through energy and motion, into a physical universe, is remarkably confirmed by the general view of scientists to-day that matter is essentially one of the forms which energy assumes. Energy may take the form of light, of heat, of sound, of electric currents, of moving bodies, of any radiation, of strains. In all of these it is still energy. We can make it as solid as we will. Still it is one of the forms of energy. Solidity is an impact made by energy upon our senses. In a very simple way we test this by experience. There before you is a body which we call a solid. You see it as a solid. That means that light rays, waves, energy, from it come with an impact upon the retina of your eye. Now you touch and feel this solid object. That means that whatever the object is composed of makes impact upon your hand as your hand makes impact upon the object, and what you feel is the impact. Matter is the name which we give to the tangible, audible, visible ways in. which energy makes impact upon the mind through what we call the senses. Greater solidity is simply a greater proportion of particles of energy, of numbers of electrons in each atom, making impact upon your senses. This is the whole tendency of "matter" as we go further in its analysis. The further we go, the more we find matter to consist essentially of energy. It consists, as men of science agree, of atoms. There are millions of atoms in the smallest visible particles of matter. They are moving at a tremendous rate of speed. The substance clearly grows less with analysis, the motion and energy

increase. What are atoms? They consist largely of space, but within that space in each atom is a whirling galaxy of electrons. There are not many electrons in each atom. They are so small that it is estimated that one of them is not more than one hundred trillionth of the size of the atom. But they move with inconceivable energy and speed. Their speed is reckoned to be at least twelve thousand miles a second. They revolve, we are told, around the nucleus at the centre of the atom a quadrillion times a second! A particle one hundred trillionth of the atom, and revolving in the atom a quadrillion times a second! Whether these figures will stand or not, the broad principle is clear. It means that the substance ceases, the energy becomes inconceivably great. We are not surprised that the physicist says that the electron is a particle of energy, and that it has no mass except its electric field. The electron can indeed be weighed, by its impact against a screen. But this simply means that its impact can be measured, for the impact of such energy, even in such minute units, is very great,--enough, it is thought, to lift one hundred pounds one foot in one second. At least it is an impact of an incredibly minute particle of inconceivable energy. These electrons, it is believed, revolve around a proton, or group of protons, at the heart of the atom. The proton is only one-thousandth of the size of the electrons which it holds in orbits around it. But it is of yet more inconceivable energy--perhaps two thousand times the energy of an electron. The particle grows immeasurably less! The power grows yet incredibly more! The proton is a particle of positive electricity. The electron is a particle of negative electricity. Positive and negative mean simply that they are reciprocal or complementary manifestations of electricity or energy. An atom is simply a balanced number of particles or charges of electricity or energy. It is all energy. But now we come to yet infinitely smaller units, which all agree are pure energy. For the electron in its inconceivable whirl about the nucleus emits units of energy. It was said not many years ago that this was done by oscillation of the electron. Now it is said that it happens when an electron shifts from one orbit to another, or, which is somewhat the same thing, that it occurs when the electron drops from one energy level to another, or that when an electron is stopped in its motion a unit or quantum of radiation is shaken off from it. And lastly it is said by a great physicist [**] that all that we can be sure of is that these units of energy are emitted by some sort of atomic shiver or shudder, and that we do not know just how it is done. In any case, they are units of energy, in strict and constant proportion to the frequency of the vibration in which they are propelled. Because this proportion is a "constant," and because these are always units, they are called quanta. They are the smallest units of energy, which means the smallest units at all, which we now know in the physical universe. They are units of pure energy. No one thinks of them as anything but energy.

The analysis of "matter," then, by modern physics carries us far beyond substance into atomic regions where all is energy and power. And we know that if the theory of protons and electrons, which are reasoned attempts to explain the energies issuing from the atom, should be discarded, we should still find ourselves before a world of atoms whose inner secret is one of vast and primal energy. Indeed there is a tendency to regard electrons as in no sense particles, not even of energy, but simply as waves or impulses of energy. There are some who see atoms now as spheroid charges of electricity or energy, units of pure power.

(Indeed, since the foregoing words were written, in 1925, based on the logic of triunity, a large number of scientists have come to regard matter as consisting wholly of energy or power in motion, in a universe of impulses, of interwoven innumerable waves of power.)

At least we know that if atoms and electrons and protons remain to us, and are analyzed into the universe of yet infinitely smaller particles moving at yet immeasurably greater speed, of which many scientists think that each electron, at least, may consist, we shall be yet immeasurably further from substance, and yet immeasurably more in the presence of pure and apparently infinite energy and power. Or if the exact forms of atomic structure which science now pictures should be altered by further discovery, the whole tendency of discovery leads us to be sure in that case that we shall be yet more in the presence of immeasurable energy passing through inconceivable motion into the field of our senses. The whole tendency of modern physics leads us where we see on every hand omnipresent power and energy passing into a tangible universe of immeasurable motion. It is exactly as the principle of Triunity in the physical universe presents it to us. It is omnipresent, primal, outspread power passing through energy into physical motion which includes both the activities and the substance of the tangible world.

How "Ether" Leads to This New Reality

What, someone may ask, becomes of the scientific idea of "ether?" Does not this conception of space as the outspreading of omnipresent Divine power, which emerges through energy into a universe of physical motion, conflict somewhat with the conception of ether as a universal substance back of all other substances?

It is true that the whole trend of scientific thought and discovery now conflicts with the conception of ether, to a far greater extent than many scientists realize. But the idea of ether makes certain very definite demands, which cannot be disregarded.

Yet the fact of space as the outspreading of Divine power, which emerges through energy into a universe of motion, though it conflicts with the conception of ether, is confirmed in a remarkable way by the scientific demand for "ether."

For empty space in inconceivable. Ether, therefore, a marvelous substance, alive with energy, is conceived. It must fill all space. It must concentrate in protons and electrons, or whatever units of energy we may ultimately find. It must then account for discontinuity, these discontinuous, separate units of energy.

Ether seems needed also to provide for continuity in the physical universe. It must provide in space between those concentrations of energy, between atoms, between protons and electrons, a medium for the transmission of vibrations or waves or quanta issuing from the atoms, or for waves or impulses without atoms or electrons, if that should prove to be the reality of things. Clearly it must provide a medium for interaction between such concentrations of energy as protons and electrons may be, and between atoms, and surely between suns and planets, and between stellar systems. By whatever name we call it, can energy carry across absolute emptiness? A medium such as ether seems needed both for electric attraction across atomic space and for gravitational attraction across solar and stellar space. It seems needed to account for all these rays, vibrations, waves or quanta of radiated energy at the one universal speed which we know best as the speed of light.

Even the upholders of the extreme New Science, the most earnest disciples of Relativity, although they discard absolute motion, or motion with reference to the ether, regard ether as still a necessity. A leading and brilliant exponent [**] of the new theories declares, for instance,--"This does not mean that the ether is abolished. We need an ether. The physical world is not to be analyzed into isolated particles of matter or electricity with featureless interspace. We postulate ether to bear the characters of the interspace as we postulate matter or electricity to bear the characters of the particles." "Characters such as mass and rigidity which we meet with in matter will naturally be absent in ether; but the ether will have new and definite characters of its own." "The ether itself is as much to the fore as it ever was, in our present scheme of the world."

The Troubles of "Ether"

But ether meets with grave difficulties. To transmit with such inconceivable "momentum" the vibrations coming to it, it must be a substance of enormous density. It must, says the most careful scientific calculation, be at least a million times as dense as lead, or platinum, or any substance which we know. This is difficult enough to grasp or to believe.

But to account, as a concentrated substance, for the "weight" of protons, ether, we are told by those who know, must be of even more incredible density. It must be dense beyond all the power of science to imagine.

But at the same time, to account for the vibrations passing through it, in some rays a million vibrations per foot per second, ether must be a substance of almost infinite elasticity or resilience. That is equally demanded of it by scientific calculation. It must have that incredible density, and then that immeasurable elasticity to overcome that density. Each is beyond belief. Together, in such incredible and almost infinitely contradictory combination, they stagger the intellect.

These are the vast and contradictory demands which science makes upon ether as a substance. They mean insuperable difficulties.

Two things, however, are clear. The factor in ether which provides for the activities of ether is its energy. The factor in ether which creates all these impossibilities of density and elasticity is its substance.

Why then imagine ether as a substance? Can we not get along with energy alone?

No. We cannot solve the problem as easily as that. If we could, ether as a substance would never have been conceived. There are great necessities to be met.

In the first place, there must be a continuity, a medium between the units or concentrations of energy, to transmit energy, whether as rays, vibrations, quanta or gravitation.

There must also be a universal reservoir of the energy which assumes all these forms, a reservoir in which energy everywhere is latent and ready to leap into these activities and to concentrate into these units.

There must also, in any living activity, whether human or Divine, be a medium between spirit and matter.

Because of these necessities, ether is imagined as a substance, to be a continuity, a reservoir, a medium, even with all the impossibilities which such a substance creates.

The Way Out

What then is the way out from the difficulty created by these demands?

The answer is not impossible. Do not the very demands reveal the way out from the difficulty? Is not the true nature of the reality which we seek as "ether" found in the very nature of these demands?

There must indeed be a medium between spirit and matter.

But there is a medium between spirit and matter.

Power radiating from the spirit of God or of man is always the medium between that spirit and all matter with which it deals.

There must be a universal reservoir of energy, the energy which concentrates itself into protons or electrons or whatever units of energy we may find.

But in a universe with God in it Creative power is the everywhere resent reservoir of energy, the source from which latent energy leaps into activity and concentrations everywhere in the universe.

There must also be a continuity, a medium, between units of energy.

But such omnipresent Creative power itself, everywhere, outspread, the source of those units of energy, is also the natural and logical medium between those units of energy which it produces. Power, which produces protons, electrons, atoms and worlds, and which where there are no protons or electrons, no atoms, no worlds, stands waiting, silent, invisible, but everywhere ready to leap into instant energy and action and to transmit vibrations, rays, quanta, gravitation, is the perfect basis and medium.

Careful consideration will show that such power provides all that we can seek in the idea of ether; it provides much that we cannot seek in the idea of ether; and it is free from all the objections to ether.

Discontinuity

For such universal power, passing into energy and motion, provides beyond question for the discontinuity seen in protons and electrons. They are normally and inevitably seen as concentrations of power and of the energy into which it passes. They are essentially particles of energy. No other qualities have been found in them. For the weight of protons is simply the impact of their enormous energy. The mass of a proton or electron is simply the mass of its electric field. Even if we give these particles substance, that substance is simply a form which their concentrated energy assumes. They are concentrations of the power which produces that energy. And no impossible "density" is needed in concentrated power as the stuff of which protons and electrons are made.

Continuity, Transmission

Such primal power provides equally for continuity. Self-evidently it provides for continuity between units of power, of which we now see everything physical to be essentially composed. What else but universal latent power could provide continuity between units of power? It provides for transmission of rays, vibrations, quanta, from these units of power. Everywhere, silent, alert, ready in an instant to emerge into energy and action, and to transmit resistlessly the vibrations or radiations presenting themselves for passage through space, it provides the perfect medium.

Momentum

It provides "momentum" which theorists ask of ether as a medium for the transmission at such inconceivable speed of the vibrations coming to it. Tremendous momentum is needed. It is hard to find it in ether. But primal outspread power passing into energy provides the momentum. For it provides the perfect momentum of resistless power leaping into life at the touch of those vibrations.

Such momentum as this could be explained, in ether, only by an incredible density, a density so great, in order to transmit such vibrations, as to exceed the density of lead one thousand times over. But no such impossible and contradictory density is needed for such momentum provided by power leaping into life.

Magnetism

Or if we call it not momentum but magnetism, as so many scientists do to-day, yet more such universal, everywhere-present power, passing instantly into energy, is the perfect medium for the transmission of such electric momentum. It provides self-evidently, as we have said, the continuity of power between units or bodies of power or energy, whether it is the continuity of attraction between protons and electrons, or the continuity of cohesion between atoms, or the continuity of gravitation between worlds.

Vibrations

Such primal power accounts too for vibrations. If resilience or elasticity is needed to account for vibrations, such universal power provides perfectly the infinite resilience, the instantaneous, infinitesimal, all-powerful elasticity, of pure power, ready at once to resist and to transmit with immeasurable energy.

Quanta

Or if quanta are seen as units of radiated energy, and as doing away with the need of elasticity, such universal everywhere-present power provides for transmitting quanta unchanged and unchecked, and indeed at added speed.

Electricity

Or if electricity takes the place of elasticity, such universal outspread power is the perfect medium for the transmission of such electric vibrancy.

Electro-magnetism

Above all, such outspread omnipresent power, with its instant energy, provides remarkably and perfectly for the universal interplay of electricity and magnetism which we call electro-magnetism, and which now begins to seem to many scientific thinkers the secret of all the phenomena of the physical universe. For electro-magnetism would substitute universal energy for density and elasticity of ether. And outspread power as the basis of the physical universe provides for that universal electric energy.

The Universal Energy

Indeed, such basic, omnipresent power agrees most strikingly with that equivalence of all forms of energy which more and more is being demonstrated. Electricity and magnetism are seen, then, to be forms which that outspread power, with its universal energy, assumes. Positive and negative energy, in protons and electrons, are reciprocal forms which that omnipresent power assumes. Quanta or waves are radiations or vibrations of energy from that outspread power in the atom. Electromagnetism is the unity, the reciprocal action, of that one interlocking, indivisible outspread power. Attraction, cohesion, gravitation, in no way explained by ether, are self-evidently in some way the action of the unity of that outspread power. The universal velocity of all rays and vibrations, the speed of light, is the standard speed at which the omnipresent creative power of God emerges unhindered into energy and action everywhere in the universe. Concentrated in protons and electrons, it slows down from its free velocity, the speed of light, not to a lesser energy, but to a less unfettered speed. That speed of light is not the norm and basis of the physical universe, as Relativity would have it. But it is the speed at which the outspread power of God,--the reality of space, and the true norm and basis of the universe,--passes everywhere into energy and action. In that sublime unvaried rate at which the Divine creative power emerges throughout the universe into unhindered action, and which we know as the speed of light, we find an added meaning for the phrase: "God is light."

The True Medium

That outspread power is the true medium between spirit and matter. Ether could not be the medium between spirit and matter. No substance can be such a medium. However tenuous, it is still substance, not spirit. The true bridge between spirit and physical motion must belong both to the spirit and to physical motion. There is only one such bridge. Power alone is such a bridge. Power is of the spirit. Power can also move. The omnipresent power of God is both power to think and power to move. It comes from the spirit of God and it passes into physical energy and motion. It is the medium between spirit and matter, between the spirit of man and the physical world, between thought and motion, between Creative Spirit and the physical universe.

The Continuum

The outspread power of God is discontinuity and continuity at once. It is source and medium at once. It provides for electricity and magnetism at once. It is spirit and matter at once. Is not this the true continuum? Is it not that primal reality toward which the idea of "ether" points? It does all that ether ought to do. It does what ether cannot do. It is free from all the objections to ether. It does not need "density." It has absolute elasticity. It has absolute penetrability. We know that energy is the essential element in "ether." Substance creates all the difficulties of ether. This universal creative power, passing into energy, and free from the impossibilities of substance, seems to be what ether really means. Is the remarkable idea of "ether" a vision, from the experimental side, of this omnipresent creative power? Witness the eloquent description of ether by a recent distinguished writer. [**] "Ether is not to be explained in terms of matter." Ether "has been spoken of as Absolute Space!!" "An electric charge must be composed of it." "Ether is the seat of prodigious energies,--energies beyond anything as yet accessible to man. All we know of energy is but the faint trace or shadow or overflow of its mighty being!" "Hidden away in its constitution is a fundamental and absolute speed." These are terms applicable only to the omnipresent power of God. They are as though chosen expressly to describe that outspread creative power which is indeed "Absolute Space," which is "the seat of prodigious energies," and which passes into a universal energy which is "but the faint trace or shadow or overflow of its mighty being."

Why then has "ether" meant substance, with all its impossibilities of density, elasticity, and penetrability? Because energy and activity cannot carry across empty space. But space which is the outspreading of power is not empty. It is full of power. That is its nature. That omnipresent power is seen as the evident source of energy and motion, of protons and electrons, of atoms, planets and suns. It is equally the evident continuum, the universal underlying medium, existing everywhere as the reality of space, between worlds, between atoms, between protons and electrons,--silent invisible, but ready everywhere at an infinitesimal touch to leap into resistless energy and activity and transmit vibrations, rays, attraction, gravitation, electricity, magnetism, quanta. This is the reality which the principle of Triunity reveals. It is the reality to which "ether" points. It is the reality toward which the whole trend of modern discovery moves. It is the reality of space. It is the bridge between the Creator and the physical universe. By it we see His creative power outspread into the dimensions which we know as space, and passing, through energy, into motion, and so into all the phenomena of a physical universe.

Particles versus Waves

A test of a supreme principle is that it should solve the problems which confront it. A riddle of science needing much to be read is one now involving the whole nature of light and of matter. Two lines of experiment seem deeply in conflict. One of the world's best authorities, the President of the British Association for the Advancement of Science, put the problem very clearly the other day,--"The nineteenth century theory of radiation asks us to look on light as a series of waves in an all-pervading ether. The theory has been marvelously successful, and the great advances of nineteenth century physics were largely based upon it. It can satisfy the fundamental tests of all theories, for it can predict the occurrence of effects which can be tested by experiment and found to be correct. There is no question of its truth in the ordinary sense. In the last twenty or thirty years a vast field of optical research has been opened up, and among the curious things we have found is the fact that light has the properties of a stream of very minute particles. Only on that hypothesis can many fundamental facts be explained. A wave theory is of no use in the newer field. How are the two views to be reconciled? How can anything be at once a wave and a particle?"

Experiment shows for instance that if a ray of light strikes a piece of aluminum, an electron of aluminum exactly equal in energy to that ray of light is displaced, and is sent out from the aluminum. That seems to indicate, exactly, particles of light composed of actual substances, to displace the particles of aluminum.

On the other hand, experiment shows that light rays collide and pass through each other. Particles could not do that, but waves could do exactly that.

"But how," says the President of the British Association, "can anything be at once a wave and a particle?" "How are the two views," involving the whole nature of light and of matter, "to be reconciled?"

"We are here face to face," he declares, "with a strange problem. We know that there must be a reconcilement of our contradictory experiments; it is surely our conceptions of truth which are at fault, though each conception seems valid and proved. There must be a truth which is greater than any of our descriptions of it."

Is not the principle of Triunity, the almighty formula of the universe, exactly such a greater truth? May it not reconcile these two sets of facts?

Is light "a series of waves in an all-pervading ether," or is it "a stream of very minute particles?"

If, as we see it by the principle of triunity in the physical universe, we have, not ether, nor some substance, but omnipresent power passing through energy into visible, tangible motion, then it makes little difference whether we call the light a stream of particles or a series of waves. It is, if you will, a stream of particles or concentrations of that universal energy; but those particles move in a universal sea of that same energy. Or it is, if you will, a series of waves in that all-pervading energy; but those waves are units, in that they are concentrations of that energy of which they are a part.

The light-ray meets a certain metal. The light-ray consists of visible motion of the energy which is the leaping into life, the embodiment, of omnipresent power. The metal also consists of tangible motion of that energy which is the embodiment of omnipresent power. Why should not the particle or wave or concentration of energy in the light-ray displace an equal concentration or electron of energy in the metal? Or, on the other hand, the light-ray meets another light-ray. The two streams of concentrations of energy collide. Why should they not pass through each other, since they are not crude "substance," but energy?

Is it not,--declares that "greater truth" of Triunity in the physical world,--in either case the same great thing with which we deal in our experiments? It is the same energy, arising from omnipresent power and becoming visible, tangible motion. When the light-ray is displacing particles from aluminum, we call the light-ray a stream of particles. When the light-ray is passing through another light-ray, we call it a series of waves. But whether called waves, or particles, or something else, this energy emerging from omnipresent power into motion is in either activity of waves or particles the same great thing, the basis and universal reality of the physical universe.

Space and Motion

Space, then, seen in the light of triunity as the outspreading of omnipresent power into dimensions, is the only possible thing, as opposed to that purely imaginary space which is the dimensions of nothing at all. It is also the inevitable outcome of modern discovery, in which the motion and energy grow infinite and the substance disappears. It is also the reality back of the conception of ether, having all the qualities required in ether, and none of the defects of ether. It is also as a supreme principle a ready solvent of the apparent contradictions between waves and particles.

All of these things confirm space as being truly the source of matter or motion, according to the triune formula, and as the beginning of the physical universe.

Radiating from the will of God, being the outspreading of His omnipresent power, space is the basis and source of energy, motion and phenomena. And conversely matter,--or energy, motion and phenomena,--is equally clearly the embodiment, the leaping into action, of that outspreading of power which we call space. Like motion, the second element in matter itself,-- like the present, the second element in time,--like the person, the second element in nature, person and personality,--or like the Second Person in the Three in One,--matter, the second element in space, matter and time, is the element which does things. It is the executive factor in the three. It is the element which experiences things. It disappears and dissolves only to reappear in triumph again and again. Nothing can destroy it. It lives, and dies, and lives again, in one unchanging round, one never-ending image of the Son in the Three in One who lived and died and lives again.

Footnotes

^118:* Robert Millikan.

^121:* A. S. Eddington.

^129:* Sir Oliver Lodge

III. THE SECRET OF THE UNIVERSE AND THE PROBLEM OF SPACE, MOTION AND TIME

How Time comes in the physical world--The organic relation of space, matter and time--What Time does--The triunity of the three--How Time connects the mind with the universe--The true and universal Relativity--The Vast Outline of the Universe--Why we can comprehend the universe--A real universe.

There remains a third set of relations with the third element in this triunity.

Time is clearly the third element. But what is time? What is the essential nature of time in the physical universe?

Time, of course, has many aspects. But essentially time is consecutiveness or successiveness. In eternity things may be, in a way which we cannot really comprehend, largely simultaneous. But here in this time-world, the world which we know so well, all things,--thoughts, motions or actions,--are one after another. They are successive or consecutive. And time is essentially that successiveness or consecutiveness.

In the physical universe time is the successiveness of motion in space. Motion in space occupies one location after another. In that is consecutiveness, or successiveness, or time. For each successive location of the motion is later than those before it. Time, then, proceeds from motion. Time may equally be called the successiveness of the locations of motion in space. Time is not a thing. Therefore, some say, it inheres entirely in the mind. But that is obviously not true. It does inhere in the mind, which is successive in its thinking, but it inheres also in motion. It is the successiveness of motion in space. If Kant could have known matter not as "objects" but as motion, he would have known that space and time not only inhere in the mind but that they also inhere in motion. Motion involves space. It proceeds from space. Successiveness, which means time, is inevitable in motion. It is the inevitable outcome of motion. Space produces motion, then, and space and motion produce time.

This is what the principle of Triunity reveals very clearly. It reveals space, through energy, as producing motion, and space and motion as producing time. The principle of Triunity, as we have seen it working in the original Source, and in the various reflections, leads to further analysis of the relations of space, and of matter or motion, to time.

For in the Three in One the Spirit proceeds from the Father, through the Son. That is the very definite presentation of them. So, in the universal reflections of triunity in matter, phenomena come from energy, through motion. So, in the reflection in the triunity of time itself, the past comes from the future, through the present. And in the wonderful triunity in man the personality comes from the nature, through the person. So now, in the combined triunity of space, matter or motion, and time, we find that time proceeds from motion, or from space through motion. It is the direct result of motion in space. It is the result of the emergence of space, or outspread power, through energy, into motion.

But the relations revealed to us in this way are still more remarkable. Time is invisible and inaudible. Yet, invisible and inaudible as it is, it is time which reveals motion, and reveals space through motion, and makes motion and space visible and audible. As in the Divine Triunity Father and Son touch us and influence us and are revealed to us by the invisible and inaudible Spirit, so in the physical universe, in vast reflection of the Trinity, space and matter touch us and influence us and become visible and audible to us entirely through invisible and inaudible time. It cannot be questioned. The vibrations of matter or motion touch and influence us and become visible and audible to our senses through consecutiveness, through successiveness of impact, that is, through sensations of time. The differences of colour or of sound are differences purely in the time of the vibrations. They are differences, indeed, of space length. But they strike us, and strike other objects, as differences in time length. The differences in wave lengths or vibrations reach us or other objects which they touch purely as faster or slower. A long wave is to us a slow one. A short wave is to us a quick one. They come to us as variations in time. So that both space and matter become real to us and affect us, and affect other things, through time. In the Trinity the Spirit, Himself mysterious, unseen and unheard by us, reveals the Son, and through Him the Father, to us. So, in the physical world, time, itself mysterious, unseen and unheard by us, reveals matter or motion to us, and through that reveals space. Invisible and inaudible itself, time alone makes the world of space and of matter or motion visible and audible to us. How closely bound together space, matter and time prove to be! Space, the omnipresent outspreading of Creative power, emerges into energy. Energy passes into motion. Motion becomes phenomena, waves of sound, of light, of colour, of all the infinite variety of the universe. Phenomena consist of successive vibrations, fast or slow. That successiveness, that fastness or slowness, is time. Everything we see, or hear, or touch, is known to us through the time length of its vibrations. So closely are space, matter and time bound to each other.

We may put this relationship very clearly in terms of motion. Potential motion is space. Actual motion is the tangible universe. Successive motion is time.

The Triunity of the Three

In all these things, the principle of Triunity points out both the absolute threeness and the absolute oneness of space, motion and time. The three are so much three that no one of the three can exist without the other two. For space, potential activity, comes into full existence only in actual motion; and this motion exists inevitably as successiveness, which is time. Space then is completely real only as motion and as time. Secondly, matter or motion is of course that potential activity of space realized. It cannot exist except as the embodiment of space. And on the other hand motion exists as successiveness, or time. Motion without successiveness is impossible. Thirdly, time in turn exists only as space comes into motion and motion into successiveness. Time in the physical world cannot exist except as the result of space and motion.

Each of the three, space, matter or motion, and time, then necessitates the other two. It is absolute threeness. The principle of Triunity in the same way points out the oneness of the triunity of space, motion and time.

Each one of the three is itself the whole. For the physical universe is all of it space, the outspreading of power, realized in motion and in successiveness. It is also all of it matter or motion, embodying space, and existing as successiveness. It is also all of it time,--space and motion acting in the form of successiveness.

It is absolute threeness, then, and absolute oneness. It is a triuniverse.

These three are of course three modes of being. They are three things which the physical universe is.

Yet more wonderfully and mysteriously we find the Triunity of God explaining the nature and working of time. In the Triune God the Spirit enters the very being of our souls, and not only influences them, but is a part of their very life. And this leads us to a remarkable thing in the nature of time. In the triune universe time enters the being of the soul and becomes a part of its very life. Neither space nor motion so enters the soul. They stay outside the windows of the senses. Neither of them becomes a part of the very existence of the mind. It is possible to think thoughts which have nothing to do with space or motion. But the mind can do nothing at all without consecutiveness, succession of thoughts. Time belongs to the mind as well as to the physical universe. As the soul was created to have the Spirit of God dwelling in it, and that Spirit, who is also the third in the Three in One, works to bring the soul to know Father and Son, so also the soul was created with time existing in it as a part of its consecutive inner life, and that element of time, which is also in the physical universe as the third great element there, works to bring the mind to know space and motion.

So the Triunity of Father, Son and Holy Spirit, whose reflection is the universal triunity, reveals one after another the subtle but absolute relations which space, matter and time bear to each other. Excepting three centres of personal consciousness in one Being, not one of the Triune distinctions is found in God which does not reveal its counterpart in the triune relations of space, matter and time, as we have seen them.

The True and Universal Relativity

The great fact of relativity has long been neglected. Now it is being felt everywhere in the world of thought. Its hour has come. No longer to-day can we think of things in the physical universe by absolute and rigid formulae. The great fact that all things of sense and space and time exist in relation to each other has found in recent years its new and right emphasis, and its priests and prophets.

It is true that those who have rescued the fact of relativity from neglect now make too much of it. They go to the other extreme. That is always done when a truth has been ignored and is again affirmed. One need not adopt every extreme speculative conclusion which has been linked with the new emphasis upon relativity, in order to give the fact of relativity its due. Nor need one make relativity the basis of all beliefs. It is only one among other facts of the universe.

But there is indeed relativity. We have seen it. We have seen it remarkable and universal in the relationship of space, motion and time in their infinite triunity. Everything in each of the three in that absolute triunity exists in relation to the other two. Nothing in any one of the three in that triunity can be computed except in terms of the other two. Space is abstract, unmeasured, unrevealed, but for motion, or matter, and time. It comes into full reality only when it is embodied in matter or motion, and consecutive motion from point to point in space, which means time. As for motion, it can be stated in the abstract as energy. But that energy is meaningless, its rapidity, and therefore its impact, cannot be measured or stated, except by space and time, the length or shortness of time required for the motion to cover a certain space, or the length or shortness of space which it covers in a given time; or the space-length and the time-length of its vibrations. And time,--this too exists in the abstract as consecutiveness; but it becomes real and measurable in the physical universe only as it results from the combination of space and motion. A unit of time is the piece of consecutiveness resulting when a certain motion of a pendulum goes a certain distance, or as a certain star moves from one point in the universe to another. Time in the physical universe is especially relative. For the momentum of the pendulum grows less. The elasticity of the spring decreases. The star moves in a universe not static but shifting in a million million orbits. Space and motion and time are to us wholly relative to each other, and we can know no one of them except in terms of the other two.

Professor Einstein reasons that since motion is all that we know in the physical universe,--space and time being intangible,--and since motion is relative, therefore all that we know in the world of sense and substance is purely relative. And since we have no absolute standards of space and time, therefore, he reasons, there is nothing definite in the physical universe.

This is a healthy scepticism. It does away with our scientific arrogance. We do not get and we cannot get absolute accuracy in our scientific measurements. We can only get almost or relative certainty.

But this which is a healthy scepticism cannot itself be made an absolute view of things. You need not sit down under the juniper tree, or throw away your test-tube, or go out and riot around in daily life as though there were no standards for you, no meaning in the universe, and no God. It is true that everything in space or motion or time is relative to everything else in each of the other two. But this does not mean that the world is one vast shifting irregularity, one infinite indefiniteness. Very much the contrary. There is an exceedingly definite thing in all this shifting interplay of space, motion and time. It is Triunity. This universal triunity, which is the reflection of the Divine Triunity, reveals to us those relations of space and matter and time. All this interplay of relativity is but the omnipresent, living, constant outworking of that absolute triunity which is the structure of the physical universe. It is, of course, not a rigid structure, but a living one, the image not of a rigid but of a living God. All of this relativity, this shifting and partially elusive interdependence, is the absolute operation of an absolute, never-changing, never-failing, universal triunity, the reflection and immanent working of an absolute, glorious Triune God.

The Vast Outline of the Universe

From this Relativity, this Triunity of Space, Matter and Time, a vast Outline of the universe dawns upon us. It is an infinite circuit going out from the life and outspread power of God, through the physical universe, and back into the eternity and life of God.

This is the outline:

The power of God is spiritual. It also from eternity spreads out into dimensions. That is space.

Acting as energy, or power to move, that outspread power proceeds, as God wills, into actual motion. Then we have a universe of motion.

The motion works out in phenomena. Then we have light, heat, colour, sound and substance.

Both the motion and the phenomena mean and are successiveness. That successiveness, of the locations of motion, of the impacts of vibration in phenomena, is time in the physical universe.

But time, like space, is both physical and spiritual. It is successiveness of mind as well as of motion. As space issues from the mind and power of God into physical dimensions, so time passes again from physical successiveness into mental successiveness, and so into the eternal mind of God.

This is the vast outline. From the all-powerful mind of God, through space, through energy, motion and phenomena, through time, back into the eternal mind of God. Out from the omnipotence of the Creator, back into the eternity of the Creator.

It is a continuous process, a vast unchanging Circuit. The universe is forever emerging, and forever returning, through space, through motion, through time, back into its Source.

"For of Him and through Him and unto Him are all things."

Why We Can Comprehend the Universe

For this same reason, because space at the source of the physical universe and time at the exit of it are both of them of the mind as well as of the physical world, the mind of man can grasp and comprehend the universe. Man can comprehend the universe, not in its extent, but in its nature.

For the comprehension of space is possible to us because space is of the mind as well as of the physical world.

And the comprehension of time is possible to us because time is of the mind as well as of the physical world.

And because motion comes out of space and passes into time, both of which we grasp, we can comprehend motion. We can comprehend it in terms of space and time. We define it as a motion of so many miles of space in so many moments of time. We can do this because space and time blend into our minds as they do into God's, and because the triune space-motion-time universe issues from the mind of God into reality, and returns again to that mind of God as its eternal goal, and the mind of man reflects the mind of God. That is why we can comprehend the universe.

A Real Universe

It is necessary to be very clear about what all this means. It is not pantheism. It is not a vision of the physical universe as a mere manifestation of the being of God showing itself in unreal appearances which are but parts of Him, a universe which seems to be space, matter and time, but is not. It is quite the contrary.

It is reality. It is the perception of how the mind of God through its omnipresent, outspread power projects into reality a universe which is not mind, and is not God, but is physical motion, and of how that universe of physical reality returns through phenomena and time into the mental eternity of God.

The physical universe becomes, then, a vast circuit from the mind of God to the mind of God, and this vast circuit is in the absolute likeness of the Three in One.

Space, or the outspreading of power, is the source, like the Father in the Three in One.

Motion or matter is the visible, active embodiment, like the Son in the Three in One.

Time is from space, through matter, as the Spirit is from the Father through the Son in the Three in One.

And as the Spirit is the return of the Godhead again into the life of God, so time is the return of the physical universe into the life of God.

"For of Him and through Him and unto Him are all things."

IV. THE SECRET OF THE UNIVERSE AND THE PROBLEM OF RELATIVITY

What is the Fourth Dimension?--Is time the fourth dimension?--The principle of the "light-year"--The Multiplication Table of Space, Motion and Time--The formula of the physicist--What you cannot do with space, motion and time--The Absolute Relationships of Space, Motion and Time--In how many dimensions do events occur?--The Formula of the True Relativity.

What is the fourth dimension?

This is not to-day a fantastic question, an idle puzzle. It has never really been so. An eager and deep-seated instinct has asked the question and great significances hover around it.

What is the Fourth Dimension?

What is that fourth dimension, which apparently does not exist, but to which reason strongly leads? It comes from motion in space, and its logic is beyond question. We may easily follow the steps of the standard argument:

The motion of a point generates one dimension, or a line, an unbroken series of locations or points. That is very clear. A moving point creates a line.

The second step is equally evident: The motion of a line or series of points generates a plane, or two dimensions. We can see this also without difficulty. A line moving sideways creates a second dimension.

And the third step is equally clear: The motion of a plane generates a solid, or three dimensions. This also we easily realize. A plane, of two dimensions, held level, for instance, and moved up and down, creates a third dimension.

But now we come to ground without a chart. What does the motion of a solid generate? Many say "A fourth dimension." That seems to follow out the logic of the first three steps. But no one has ever experienced a fourth dimension.

Does logic truly run away from reality in this matter? It seems to. Indeed, once started on that way, it seems beyond control. If a fourth dimension, why not a fifth? What would the motion of a four-dimensional figure generate, except a fifth dimension? And why not a sixth and seventh and eighth dimension? Can it be that logic runs out into an endless chain of such dimensions, further and further from reality? Can there really be such a divorce of logic from reality, and of logic based not on uncertain premises, but on the three dimensions of space and of geometry? How can the threefold reality of geometry so suddenly land us, at one move, in a world of ever-increasing unrealities?

Yet surely the motion of three dimensions generates something. The motion, the reality, of one dimension, a line, generates a second dimension. The motion, the reality, of two dimensions, a plane, makes three dimensions. What further reality do three dimensions generate? What is the product of three dimensions? Or if there are other dimensions also, what is the product of all possible dimensions?

Is Time the Fourth Dimension?

That fourth property or "fourth dimension," or, if there are more dimensions, the final property, of space is not time.

Einstein, who roams much in the realm of the fourth dimension, regards time as the "fourth dimension," because, he says, things happen in three dimensions of space and in one of time. Indisputably things do happen in time as well as in space. Einstein believes that this fourth property or "fourth dimension," or time, is the "continuum," the thing which binds everything, including the dimensions of space, together. A point in "four dimensions," or a thing which happens, he rightly calls an "event," because it must be more than a point, it must be something which happens, in order to have a place in time. And he is sure that things occur in three dimensions of space and in one of time. Equally sure are many of his followers. Indeed much of the fabric of the newest scientific view of the universe, with its overthrow of classical physics, is built up on the assumption that, because things happen in three dimensions of space and also in time, this is a four-dimensional world, with three dimensions of space and one of time. And many announcements in astronomy and physics, scientific in their working out and terminology, are in reality based upon the purely speculative foundation of the theory of "Space-time," and are of no more certain validity than their basic speculation.

And now it is announced that the entire physical universe becomes one single reality,--namely, space,--and that space has swallowed up time, by making time its fourth dimension.

And finally, on the other hand, a new geometry, to embrace all the facts of the physical universe, is projected, with time as a dimension of space set forth as the basis of the new system.

But is time the fourth dimension of space? All of these things depend upon that assumption. Do the evidences which are brought to demonstrate that time is the fourth dimension of space stand up against the strong wind of common sense or of reason? That is what we have a right to consider for ourselves.

Do Light-Years Make Time the Fourth Dimension?

In order to prove that time is the fourth dimension it is very much the thing to-day to point out that when one would measure vast distances in the universe one must go beyond the old space methods and use time units as well as space units of measurement. When we have a vast astronomical distance to be expressed, we call in the aid of time. We say that such a star is a million "light-years" away. We mean by a light-year the distance which light, at the rate of approximately 186,000 miles a second, goes in one year. Does not this prove that time is fundamental in great basic measurements? Does it not mean that time is an essential element of space? Does it not show that time is a "fourth dimension" of space?

At the cost of disagreeing with what many minds regard as an axiom to-day, one can only say, if one is doubtful, "It does not seem to me to prove what you I say that it proves." Far the fact is that the now common light-year measure of distance does not arise from the nature of distance or of space. It arises entirely from the limitations of our minds. Because we cannot grasp more than a certain number of smaller units of distance, we combine them into larger units for our mental convenience. It is exactly as when, to avoid `too great a number of inches, we say feet, instead of inches, or when, to avoid too great a number of feet, we say miles instead of feet. In time-measurements, also, when seconds grow too many, we say minutes, and when minutes become too many, we say hours, and when hours multiply too largely, we say weeks, and when weeks add themselves into a great total, we say years. We manufacture larger units to bring the total number better within the grasp of our minds. We manufacture light-years simply as a larger unit of measurement. If the use of time in measuring distance lay in the real nature of measurement of space, we should have to use time in all measurements of space. We should have to use it as a factor in measuring short distances. But we do not use it so at all. We do not use time as a factor in measuring feet or metres, or in measuring miles on the earth. The only people who use it so are those whose mental ability is so low that they cannot compute space distances at all, and who say, "It is so many days' journey," or "so many hours' journey," or "it is as far as a horse would travel between sunrise and sunset," or "as far as a man could walk, carrying a sheep, between moon and moon." Such things, day's journeys of a horse, or month's journeys of a man, or year's journeys of a light-ray, do not mean that time is a dimension of space. They signify simply a greater or less degree of mental inability to grasp large numbers of units of distance. Equally well the savage meets the difficulty by saying, "It is ten times as far as from the mountain range to the distant river," or the scientist by saying, "It is a million times as far as the distance from the earth to the sun," or the mathematician by saying, "It is a trillion miles carried to the 10th power." It is all a matter of constructing larger units of measurement, so as to bring down the total number of units to the range of our comprehension. It does not at all show that time is a dimension of space.

The Multiplication Table of Space, Motion and Time

But the sensible way to settle the question is not by arguing against a mistaken association of time with space. It is by working out and presenting the actual relations. It is useless to argue that time and space are not related. They are related. They are commensurate. They are indeed so commensurate that we can describe one in terms of the other. The trouble is not that this is true. The facts should never give us trouble. The trouble is that it is only a part of the truth. It is only what the triunity of space, motion and time shows of all three of those elements. They are all three commensurate. And it is only through its part in that triunity, and by its connection with space through motion, that time is commensurate with space. It is only through motion that we can measure space and time in connection with each other. We can measure space by motion and express the result in terms of time.

100 miles covered by a motion of 25 miles an hour, and divided by 25 to get the number of hours, results in an elapsed time of 4 hours. 100 / 25 = 4. Hence the basic rule:

1. Space measured or divided by motion gives time.

The distance, divided by the rate of motion at which you travel, will tell you how long, or how much time, it will take you. This is the basic rule. Space can be measured by motion in terms of time because motion touches both space and time and is expressed in terms of both. There is no direct relation between space and time. 100 miles and 4 hours have no connection unless there is motion which covers those miles in those hours and links the two together. The time is related to the space through the motion.

Now this is because motion must always touch both space and time, and be commensurate with both, and with both at once. Motion can be expressed only by both space and time at once. Space can be reckoned by itself,--e. g., "100 miles." Time can be reckoned by itself--e. g., "4 hours." But motion can be reckoned only by space and time, and only by both at once. "A rate of motion of 25 miles" means nothing. "A rate of motion of one hour" means nothing. But "a rate of motion of 25 miles an hour," both space and time at once, is well defined.

And this is the triune relationship which we have already seen,--that motion comes out of space and passes into time and so links space and time together. It is the principle of the space-motion-time universe. The creative power, whose omnipresent outspreading into dimensions is space, emerges into energy and motion, and motion issues in successiveness or time.

Time then is related to space through motion, and only through motion, and only in the triune relation of space-motion-time. We can also use time by motion, with space as its result. Using a time of 4 hours, a motion of 25 miles an hour will cover 100 miles. 4 * 25 = 100. That is, for the second rule:

2. Time multiplied by motion gives space.

The time elapsed, multiplied by the rate of motion, will give you the distance you travelled.

This is manifestly the reverse of the first rule. Because space divided by motion gives time, time multiplied by motion gives space again. The quotient multiplied by the divisor gives the original number again. That is what it means. The time multiplied by the motion does not actually produce space. Both motion and time presuppose space. It is simply an arithmetical transaction in which we multiply the quotient by the divisor to get the original dividend, the 4 by the 25 to get the 100 out of which they both came. And the connection of the time with the space, whether we are dividing the space or multiplying the time, is always through the motion. There is no connection between 100 miles and 4 hours, unless motion comes and spans the 100 miles in the 4 hours, and so binds the two together, or in turn multiplies the time to give the original space again. It is the triune formula. A third relation also issues from the first. Since "space divided by motion gives time," therefore:

3. Space divided by time gives motion.

The distance, divided by the time it took you to cover it, gives you your rate of motion. A space of 100 miles covered in 4 hours, and divided by the 4 to get the miles for one hour, shows a motion of 25 miles an hour. 100 / 4 = 25.

This does not mean that time measures and divides space as motion does. We cannot measure space by time. We cannot cover 100 miles by 4 hours. They are not commensurate, that is, they are not commensurate unless motion comes between, and connects them. But "100 miles covered in 4 hours" means "100 miles covered by motion in 4 hours." It was the motion which measured and divided the space and gave the time.

"The space divided by the time gives the motion" simply sets forth then the converse of the basic fact that "the space divided by the motion gives the time." For you can always divide the original number by the quotient to get the divisor. As 100 / 25 = 4, so 100 / 4 = 25. It is an arithmetical transaction. And it expresses again the basic fact that "space measured or divided by motion gives time,"--the triune principle of the space-motion-time universe.

 [**] (The physicist uses the constant formula in his work: v = d/t,--"the velocity = the distance / the time." Or, which means the same thing, "the rate of motion = the space / the time." Or we can put the equation in the other order: d/t = v, "the space / the time = the rate of motion."

We see now, in the light of the universal triunity, and of the triune relations of space, motion and time, the great reason which underlies this working formula of d/t = v, or the space / the time = the rate of motion. It is not because the time measures and divides the space. We cannot measure or divide space by time. Space and time are not in themselves commensurate. 100 miles has nothing to do with 4 hours, for instance,--unless motion covers that 100 miles in that 4 hours, and links the two together. Then "100 miles / a motion of 25 miles an hour gives a time of 4 hours." "The space / the rate of motion = the time." Or, in terms of the physicist's equation, d/v = t, or t = d/v. That is the basic relationship,--"space / motion gives time,"--without which there are no relationships between space, motion and time. But because d / v = t, or d/v = t, you can, if you have the distance and the time, and want to find the velocity, divide the dividend by the quotient, divide the d by the t, to get the divisor or velocity, v. Then at once you have as a working rule, based on our triune law that space / motion = time, or d/v = t, the reverse rule, dividend/quotient = divisor, or space /time = motion, or the useful tool of the physicist, that d/t = v. The triune law of all things, and of space, motion and time, gives the reason. d/t= v simply because d/v= t;--the distance divided by the time equals the speed, because the distance divided by the speed equals the time. You can always divide the distance by the time and get the velocity,--even though distance and time are not in themselves commensurate, because "the distance or space covered and divided by the velocity or rate of motion gives the time."

And a fourth rule follows as a consequence of the basic rule. Since "space measured or divided by motion gives time," then you can reverse the process, and find that:

4. "Motion multiplied by time gives space."

The rate of motion, multiplied by the time elapsed, gives you the distance.

A motion of 25 miles an hour, for a time of 4 hours, covers a space of 4 times 25 miles, or 100 miles. The motion multiplied by the time gives the space. 25 * 4 = 100. This obviously does not mean that the 100 mile space exists as the product of the 25 mile motion and the 4 hour time. What it means is that you can multiply the divisor by the quotient and get the original number or dividend. Since the space divided by the motion gives the time, then the motion multiplied by the time will give the space again. It is a matter of arithmetic. It comes from the basic fact that the space divided by the rate of motion gave the time. It expresses and shows in reverse the triune formula of space-motion-time.

These four rules all express the same relationships,--that space is always the source in this triunity,--motion is always next,--motion alone is directly commensurate with space and can measure space,--time always results when motion measures space,--motion can be measured only in terms of both time and space, which it links together, they are all three then commensurate with each other in this triunity of the three,--but time is commensurate with space and can be related to space only as a part of that triunity, as it is linked with space through motion, or comes from space through motion.

[There are no other such relations between the three. For other operations are not possible. The following paragraphs seem meaningless because to a certain extent they are meaningless. But their impossibility is instructive. Yet you may skip these "cannots" if you will.

You cannot divide motion by time.

"A rate of twenty-five miles an hour divided by four, the number of hours passed," means nothing, and gets no results.

You cannot divide motion by space.

"A rate of 25 miles an hour divided by 100, the number of miles traversed," means nothing at all. If you say "25 / 100 = 1/4 miles," it means simply that 1/100 of 25 miles an hour is 1/4 mile an hour. But it has no significance in the relations of space, motion and time, any more than the similar facts that 1/200 of 25 miles an hour = ###8539### mile an hour, or that 1/50 of 25 miles an hour = 1/2 mile an hour. It has nothing to do with one hundred as the number of miles.

And you cannot divide time by space.

"Four hours divided by 100, a certain number of miles," means nothing. You can, it is true, divide the four hours by 100, and get 1/25 of an hour, the time required to go one mile. But it means simply that we have reduced the dividend, 100 miles, in "100 / 25=4," to one mile by dividing by 100, and so as a matter of arithmetic we must divide the quotient, four hours, by 100 also, and we get 1/25 of an hour. We could equally well divide the 100 by fifty and get two miles, and the four hours by fifty and get 2/25 of an hour. We simply reduce both dividend and quotient by the same ratio, and we chose to divide the 100 miles by 100 in order to reduce the distance to one mile. In reducing the dividend and the quotient by the same ratio we have but stated anew, on a scale 1/100 as large, the formula that "the distance or space / the rate of motion gives the time."

And you cannot divide time by motion.

"Four hours divided by a rate of motion of twenty-five miles an hour" means nothing and accomplishes nothing. You can only multiply the time by the rate of motion, the quotient by the divisor, and get the dividend or distance.

And you cannot multiply space by motion.

"One hundred miles multiplied by a rate of twenty-five miles an hour," would be 2,500 miles, which is arithmetically correct, but an absurdity. You can only divide the one hundred by the twenty-five, and get four, the number of hours elapsed. You can only divide the space by the rate of motion and get the time.

Neither of course can you multiply motion by space.

"Twenty-five miles an hour multiplied by one hundred, the total number of miles traversed," would again be 2,500, this time "2,500 miles an hour," which is again absurd and meaningless.

Neither can you multiply space by time.

"One hundred miles multiplied by four, a certain numbers of hours," means nothing. You can only divide the one hundred miles by four, the number of hours passed in motion, and get the twenty-five miles an hour, the rate of motion.

Neither of course can you multiply time by space.

"Four hours multiplied by one hundred, a certain number of miles," means nothing. There is no connection, no relevancy, between the hours and the miles. It is only when the four hours are the time resulting as a certain motion covers the space of one hundred miles that you can relate the four hours and the one hundred miles to each other. They are related to each other solely through the motion. And then you can only divide the space by the time, the one hundred by the four, and get twenty-five, the miles per hour, because the space was the dividend and the time was the quotient in "the space / the rate of motion = the time."]

These include all the possible or impossible combinations of the three,--space, motion and time. Even the impossibilities emphasize the relationship which the four possible combinations so clearly express. This relationship is so self-evident as to be axiomatic:

The Axioms of Space, Motion and Time

1. Space can be measured or divided by rate of motion, with time as the result. This is the basic relation: "Space measured or divided by motion gives time."

2. Space therefore, the dividend, can as a matter of arithmetic be divided by time, the quotient, with the rate of motion, the original divisor, as the result.

3. The rate of motion and the time can be multiplied, either one by the other, the divisor multiplied by the quotient, or the quotient multiplied by the divisor, the rate of motion multiplied by the time, or the time multiplied by the rate of motion, with the dividend, the distance or space traversed by that motion in that time, discovered again by the multiplication.

4. Space is the source. It is space which is traversed and measured or divided up by the rate of motion, with time as the result.

5. Motion links space and time together; it emerges from space and issues in time; and it can be measured and expressed only in terms of both space and time.

6. Time is commensurate with space to this extent, and only to this extent, that since time is the product when motion emerges and traverses space, time is commensurate with space through motion.

Two facts stand clear in these relationships:

First, that all these relationships express the absolute formula of triunity, namely, that space is the basis, motion traverses the space, and time is the result.

Second, that time is related to space only through motion, and is not the "fourth dimension" of space. It is, rather, the third factor in the triunity of space, motion and time.

Light-Years Agree with This

Light-year measurement is in entire harmony with these axioms. The astronomer knows a certain vast distance. It is a distance so vast that the figures benumb the mind, and convey no real impression. He wants to express the distance vividly. He says "We will measure it by motion. A certain motion, the speed of light, moving at a rate of 186,000 miles a second, spends so many years, so many light-years, or years of that motion of light, in covering that space." It is that space measured or divided by that rate of motion of light, with years, or time, as a result. It is the triune formula, "Space measured or divided by motion gives time."

Or let us put it not from the astronomers' point of view but from the hearers' or readers' point of view. Here, then, is a vast distance. We do not know that distance, as the astronomer does. But we do know the speed of light. And we do know how long a year is. And so the astronomer tells us that the speed of light, multiplied by a certain number of years during which that light moves, gives the total distance over which that light has moved. It is the rate of motion which traversed the distance, multiplied by the time elapsed, to give the distance,--the motion multiplied by the time to give the space. It is the triune formula,--"The motion multiplied by the time, the divisor multiplied by the quotient, gives the original space."

It reveals time yet again, very strikingly, not as a "fourth dimension" of space, bound in direct union with space, but rather as having its sole connection with space through motion, whether it is "space divided by motion gives time" or "the motion multiplied by the time gives the original space;" and as having its only connection with space in the fact that time is the third factor in the vast triunity of space, motion and time.

In How Many Dimensions Do Events Occur?

The theory of relativity has this value, besides others, that it clearly recognizes the great fact of Motion as the essence of all physical things. It realizes that things "happen" rather than merely "exist," and are "events" instead of inert and static things, because they are actions, motions, conjunctions of forces, collisions or propulsions of electrons, with their incredible motion. And of course where motion so exists time enters as an element. For time in the physical world is the successiveness of motion in space.

But it is not correct to say that things occur in three dimensions of space and in one of time, and so to make time a fourth dimension or property of space, even though this is almost an axiom of much present-day thought. It is an evanescent point of view.

For, as a matter of fact, events do not so occur.

They occur in three dimensions of space and in three dimensions of time. Every event is in future, present and past.

It occurs in present time.

Before its occurrence in present time it is in future time.

After its in present time it is in past time. It cannot be in time without being in all these three elements of time.

As for those important events which we call thoughts, many of those, which have nothing to do with the outer world, but which are always consecutive, occur in no dimensions of space and in three dimensions of time.

But of things in the physical world, all of which, being motion, do "happen" or "occur," all of them occur in three dimensions of space and in three of time. Time is not the fourth dimension of space. Time in the physical world is, as we have seen, self-evidently the outcome of the relation of space to motion. It is the successiveness of motion in space. It is the third element in that universal triunity of space, matter and time. But it is not a fourth property added to the triunity of space. It is the third property in the greater triunity of space, matter and time. And it is itself a triunity, of future, present and past. For the True Relativity is not a vast and shifting irregularity, a systematized uncertainty. The True Relativity is a vast, invariable, regulated, infinite interplay and triunity of Space, Motion and Time. It is a Triuniverse in the reflection and image of a Triune Creator and Ground of the universe.

V. THE SECRET OF THE UNIVERSE AND THE PROBLEM OF BEING, OR EXISTENCE.

The real Fourth Dimension--In how many elements do things happen?--The mighty Principle and Process of the Universe--"Being" versus "Becoming," in the light of the Process of the Universe--The Being of God in the Light of Space.

And now we have come to the point where we can gather together certain great conclusions. They rise, range beyond range, in almighty mountain heights. They are not of our handiwork. No principle smaller than the being of God could raise such tremendous conclusions. They are not of our making, but they are as self-evident as the vast "Circuit of the Universe" we have already seen.

1. The Real Fourth Dimension

Neither Time, we have seen, nor a series of further dimensions is that reality which is the great and final product of the dimensions of space. We must look further.

What then is the reality which those dimensions produce? We know that the motion, the existence, of one dimension, a line, generates a second dimension, thus making a plane. The motion, the existence, of two dimensions, a plane, generates a third dimension. What is it which three dimensions generate or produce?

The answer, when we brush aside all theorizing, and gaze directly at the facts, is very clear. What is it which three dimensions produce? Whatever else they may produce, whatever other dimensions they may or may not generate, the supreme thing which they produce is Space. For when three dimensions are combined, space becomes a reality, and not until then. Space is the product, it is the combination, it is the unity, of the three dimensions. Space is the reality which the three dimensions generate.

If there are further dimensions, the great final reality which they combine to produce is the same space. If they did not combine in the unity and reality of space, the dimensions would possess no reality. All is imaginary until there are three dimensions. Space is the great reality which they generate. And that series of further dimensions, which mathematical speculation seems to demonstrate, is simply the ever-intensifying reality of Space.

clearly stated and thereby confirmed when we remember the yet higher way of demonstrating further dimensions. We have a pencil or sheaf of an infinite number of lines all passing through a certain point. That infinite number of radiating lines means, of course, three dimensions. This is the argument of descriptive geometry.

Now we can reason that we have an infinite number of such pencils passing through the point. That means apparently a fourth dimension. Now we have further an infinite number of such points. That means equally a fifth dimension.

This is the standard argument of higher geometry for the fourth dimension. But it is after all, we can see now, essentially an intense statement of the reality of space. We have the pencil of lines passing through the point. That means three dimensions, and the reality of space. All this infinite number of radiating lines in the pencil produces only the one simple three-dimensional reality of space. They have not carried space beyond that simple reality.

But now there are an infinite number of such pencils. We have infinitely multiplied that simple, three-dimensional reality of space! We have an infinite repetition or intensification of that reality of space. We have carried that reality into infinity.

And now there are an infinite number of such points. With these an infinite number of times we yet further reaffirm that same infinite reality of space.

The dimensions, in other words, produce an ever intensifying, an infinitely intensifying, reality of space. The so-called fourth and fifth dimensions mean the infinite reality of space itself.

The Fourth Dimension, so much sought and so much desired, is Reality. It is existence.

The Moving Reality of Space

We see now how it is that matter or motion gives reality to space.

For we have seen earlier that real space, not a vast stretch of non-existence, but real space, consists of the outspreading of Creative power, power which can move, and does move.

Now we see that it is in the motion of space at any or many of its points, first into lines, then into planes, then into three dimensions, that space has its final complete reality, and that if there is motion into yet further dimensions they simply make that moving reality of space yet more intensely real. They carry it into infinite certainty.

2. In How Many Elements Do Things Happen?

And we can answer now, in the light of greater, more absolute principle than any other which concerns the question, the query "In how many elements do things happen?" We can go far beyond the inadequate answer that they "happen in three dimensions of space and in one of time." For they happen in far more than that.

Einstein, and many others, have felt, as indeed we all) have felt, the fact that time is in some sense the outcome of space. But he did not realize that, with motion emerging from space and with time as the outcome of space through motion, we have an unbreakable triunity of space, motion and time. He did not realize that in saying that things happen or take place in three dimensions of space and one of time, and in emphasizing the fact that things "happen" instead of merely exist, because the physical world is a world of motion, we are admitting yet another element into the case, the element of motion.

For we have seen that it is only through motion that time comes from space.

Time is related to space only and wholly through motion. Time is not the direct fourth property of space. Unity, reality, as we have seen, is the fourth property of space. Time is rather the third element in that triunity which includes space and matter and time, all three triunities in one.

If we should state the full fact we should say that things happen in three elements, not in two. Not in space and time. They happen in space and motion and time. Any lesser statement ignores the modern universe of motion.

If we state the fullest fact, we should say that things in the physical universe happen not in four dimensions or elements, three of space and one of time. They happen in nine elements; three of space, height, length and breadth,--and three of matter,--energy, motion and phenomena,--and three of time,--future, present and past.

Or if we put it most clearly we should say that things happen in three triunities, and in the three combined in one great triunity of Space-Motion-Time.

And if we put it most clearly of all we should say that things in the physical universe happen or take place or exist in three triunities,--space, matter and time,--and in one great triunity of those three combined,--and that these three universal triunities, and their combined all-inclusive triunity, are the absolute image in every possible way of the supreme Triunity of Father, Son and Holy Spirit.

3. The Process of the Universe.

It is possible to see the process of the existence of the universe, in the light of God.

For a yet further universal fact is waiting for us in regard to this process of the existence of space which reflects the existence of God.

What we have just found to be true of triune space is true also of the other triunities which compose the universe.

In space one dimension generates the second. The two dimensions generate the third. The three dimensions generate the reality, the existence, of space. This is the principle of the existence of space.

Now this same principle is true of the other universal triunities. It is true of matter. Energy generates motion. That is evident. Energy and motion generate phenomena. That is equally clear. Energy, motion and phenomena combined generate or constitute the existence of matter. Matter becomes a complete existence in the union of those three elements. That is basic.

The principle is equally true of the existence of time. The future produces or generates the present; the present comes wholly from the future. That is unmistakable. The future and the present produce the past; for out of the future comes the present, and out of the present comes the past. Of that there is no question. Future, present and past combined constitute or generate the complete existence of time. That is self-evident.

The principle is true also of the triunity of human existence. His essential being or nature necessitates and produces the character, the mentality, the person, whom you know. Nature and person generate personality, for the nature, working through the person, or the person, working from his inner nature, produces the personality which touches others. And the three combined constitute complete human existence.

The principle is true also of the triunity of space--matter--time. Space, the outspreading of the power of God, generates motion or matter. Space and motion combined produce time, which is the successiveness resulting as motion traverses space. Space, motion and time combined constitute the physical universe.

We have then in this triunity the process of the existence of space, of matter, of time, of human life, and of the space-matter-time universe. The pattern is a process. It is the process of the universe.

It is always the same. The first factor produces the second. The first and second generate the third. The three generate complete existence.

This is the process of existence, whether of space, or of matter, or of time, or of the space-matter-time universe, or of human existence.

The heart would be weak and the mind would be dull indeed which could not be stirred by such a vision, so vivid, so real, so universal, of the almighty process of the universe, in the image of its Creator.

4. "Being" versus "Becoming," in the Light of the Process of the Universe

Systems of thought have from the beginning presented the question of "being" versus "becoming." Is "being" or is "becoming" the secret of existence?

Does this seem a somewhat abstract question? On the contrary, no broader division of the whole field of human thought can be found than the division which is described by "being" versus "becoming." For generations ancient philosophy vibrated between the two. The march of Greek thought swung first to one side of the road, then to the other. Successive schools of philosophy were built about one point of view or the other. To a splendid series of thinkers the universe was a fixed fabric. It was permanent, and could be studied at leisure. It arose around one as a great framework of certainty. If one could but find the framework, the formula, of absolute "being," one had the secret of all existence.

To other strong thinkers the constant "flux" which they could see in all things, the changing positions of the stars, the turns of the tides, shifting winds, fire appearing and disappearing, the fluctuations of the soul, seemed the basic fact of the world, and they sought to find a formula to explain the universe in terms of change and motion. All things were always and every moment becoming what they were, and then immediately becoming something else. "Becoming" seemed, then, the secret of existence.

Modern thought too has moved in an immense vibration, first toward the one and then toward the other position. For generations great philosophy has sought the "ideal," the "absolute," the fixed "reality," the unalterable fabric of things, the changeless principle, so that we could know absolutely, and test all things by that absolute knowledge. Marvelous things the great classic masters of modern philosophy have done with the study of "being." Science too has long sought the exact and unalterable. It has desired to be an instrument of vast precision, a sextant of certainty. Fixed facts, unbroken laws, absolute order,--these, we have felt, are the pride of science.

But now what shall we do? Philosophy and science have fallen in love with the other ideal again. We have crossed the road. The new Science, the new physics and astro-physics, see "flux" as the very Fact of the universe. Bewildering motion, nothing at rest, universal change, breakdown of atoms, free electrons, infinite variation, transition everywhere, universal "becoming" at immeasurable speeds in infinitesimal instants, these are what we see now in the universe.

"Development" is the lens through which we gaze at all things to see them as "becoming" rather than already "being."

The theory of Evolution is in all realms the most thoroughgoing theory of "becoming" as a universal formula that the world has ever known. Pragmatic philosophy tells us that truth is what becomes true to us in practice.

Ritschlianism in religion says to you that truth is what becomes true to you in experience.

Psychology, the study of how the soul acts and comes to be what we find it to be, takes the place of metaphysics, the study of what the soul unalterably and ideally is.

Behaviourism pictures human life as continually coming into existence by its acts, a continuous "becoming."

Relativity finds the world one vast continuous flux, one universal "becoming."

The Quantum theory depicts all things as existing by virtue of their incessant change.

The modern world of thought has truly veered far over from fixed "being" to endless "becoming." The whole swing of the world pendulum from being to becoming is expressed in the a. b. c. of a genuine modern philosopher, [**]--"What is the precise meaning of the word 'exist'? I find, first of all, that I pass from state to state.--I change, then, without ceasing.--The truth is that we change without ceasing, and the state itself is nothing but change."-- "Philosophy--is the study of becoming in general."

What shall we say, then?

Is the process of existence "being" or "becoming?" Is it static or in constant flux? It is no easy choice which the ages put before us. But must we really choose between the two? Is it truly a necessary antithesis? Do "being" and "becoming" so exclude each other as the basic process of the universe?

Shall we not look upon the universe in the light of its supreme, triune Reality?

Shall we not bring the riddle of "being" and "becoming" to the supreme Solvent?

If we do this, "being" and "becoming" as a process of the universe are seen not to exclude each other at all. Rather, the triune process of the universe, the universal process of existence, gathers together the principles of "being" and "becoming" in a great reconciliation.

For the universal process of existence, in which the first factor produces the second, the first and second generate the third, and the three generate complete existence, is at once both "being" and "becoming."

It is, in the first place, the process of all existence, of all "being," in God, in man, in space, in matter, in time, in space-matter-time, in everything.

And at the same time it is in all things physical and spiritual a process of existence, a mode of "being," which is in itself a constant "becoming," a constant generation of second from first, of third from first and second, and of existence from the three.

We see then the almighty reconciliation.

For existence is both "being" and "becoming," at once. The circle of being is within itself an incessant and never-ending becoming.

The mighty process of existence does more than recognize both "being" and "becoming." It is both "being" and "becoming."

It is static, for it is changeless "being."

It is endless flux, for it is constant "becoming." It is both at once.

It is a universal process of existence by which "being" is itself a constant "becoming." In this triune universe, in the image of the Triune Creator, questions of "being" and "becoming" pass away. They melt before the sunrise. They merge into one supreme reality.

This is the process of the universe. This is the universal process of existence, in absolute likeness of the Three in One, Father, Son and Holy Spirit.

And now we rise yet higher.

It is possible to see the Being of God in the light of His universe.

For the facts of Space cast a revealing light upon that absolute Existence of which Space is but the pale reflection in the outer universe.

It may be that the endless quest of mathematicians, and of other seekers after reality, the constant search for the Fourth Dimension, really gets its urgency from the deep, unconscious thirst of the soul for God. It may be no abstract curiosity, but an instinct that beyond these three dimensions lies the supreme Reality.

The vision of the Being of God which breaks upon us when we see it in the light of Space does not depend upon the facts of Space. The facts of Space simply call our first attention to that vision. Once seen, that Being of God is self-evident. It cannot be forgotten. It becomes the Reality at the heart of the universe.

The Divine Triunity is not $1 + 1 + 1$. A crude objection has sometimes regarded it so, and pointed out that $1 + 1 + 1 = 3$, not 1. It sums up the one great objection to the Trinity. How can there be unity in $1 + 1 + 1$?

A surprisingly large number of earnest and thoughtful people through the ages have been influenced by this objection. The vast majority of those who know God have not indeed been deeply moved by this objection, for they have felt instinctively that the life of God is an infinitely more vital thing than can be expressed in a crudely materialistic formula such as $1 + 1 + 1$.

But we can learn here from the intensive nature of space and its dimensions, as those dimensions combine to produce reality. Here then is the great answer, revealed by the nature of the unity of space, but self-evident in its own right:

The Trinity, the Divine Triunity, is not $1 + 1 + 1$. It is no more so than space is. Space is not height + length + breadth. That would be a childish conception of space. Space is too immaterial a thing for such a crudely materialistic formula. Add two dimensions together, and you do not get the area of a square.

Add three dimensions together, and you do not get the contents, the total space of a cube. Space is not height + length + breadth. Everybody knows this. When you have three dimensions you never add them. That is meaningless. You multiply them. Space is height * length * breadth.

You multiply the three dimensions, and you get the contents, the space, enclosed by the three dimensions. Height * length * breadth = space. Until you have the three dimensions, multiplied by each other, you have no space. You have only an imaginary line, or an imaginary plane. Space and reality come when you multiply the three dimensions. Height * length * breadth = space and reality.

So is the Divine Triunity. The Trinity is not 1 + 1 + 1. That is a childish conception of it. The Trinity is an even more immaterial thing than space. If it is anything at all, it is life. It is the life of God. It is not like adding blocks of wood to each other. We have put away childish things. We are dealing with life, and Divine life. The Trinity is not 1 + 1 + 1. No such crudely material conception means anything in connection with it. The Trinity is 1 * 1 * 1. That is life. 1 + 1 + 1 = 3. But 1 * 1 * 1 = 1. That is God.

The Three in One as brought to us in the God of the Bible and of the universe of space, matter, time and men is that kind of absolute unity in which each of the three is the whole. Each is not a part of God. Each is God. Each is the whole. We have seen such Unity reflected, infinite, Divine and spiritual though it is,--in space, in matter, in time, in man, and in the space-matter-time universe. May we now see its mighty significance, not reflected, but in Itself?

In 1 + 1 + 1 each is a part of the whole. Each is one-third of the whole. But in 1 * 1 * 1 each is the whole! For in such multiplication each unit multiplies and permeates every part of the whole. Each is most intensively the whole, and every part of the whole. The Trinity is not an inert division of God into three parts. It is not 1 + 1 + 1. It is life. It is 1 * 1 * 1. It is One * One * One = One. It is multiplied, infinitely intensified Reality.

It is a living, active, intensive mode of being, in which each of the Three interacts, penetrates, intensifies, lives in the other Two, and each is the Whole.

One * One * One produces an intensive, multiplied Unity, deeper, greater, more One, than simple Unity could be.

God is more deeply, infinitely One than He could ever be if He were not also Three.

O God, we adore Thee! O God, Thou art Life! Forgive us if we have ever talked of Thee as though Thou wert material, or arithmetical, or anything less than infinite, immortal Life; and have tried to measure Thee, who art Life, by our little formulae; and doubted Thee, because one stone plus one stone plus one stone makes three, not one, when Thou art not stones but Life. We adore Thee, who art Life, and art infinitely more One because Thou art forever Three.

^172: Bergson, "Creative Evolution."*

VI. THE SECRET OF THE UNIVERSE AND THE PROBLEM OF CHANGE AND PROGRESS

Change and changelessness in the universe--What is the law of progress and fixity?--Evolution not a universal law of progress--The Law of Changeless Change, or the Universal Law of Progress--The Law of Progress practically applied--Laboratory proofs of the Law of Progress.

It is certain that the universe is one of continuous change. No observer can hesitate to admit what is so evident to all. Development never ceases its endless round of operations.

It is equally certain that the universe is one of marvelous fixity. Forever changing, it is forever the same.

Its constant change and its unending changelessness impress the mind with equal astonishment. Its universal whirl of change bewilders us. Its age-long changelessness mocks us or encourages us, as the case may be. Thinking has always vibrated between changelessness and change, between fixity and progress, as the law of the universe. Entire schools of thought have been based on one or the other view. The vibration from one to the other is a matter of emphasis. Everyone of candid mind must recognize both facts.

For many centuries, until the last century, the great fact of change, of flux, as the ancients called it, has not had its fair share of emphasis. The world has been regarded too much as a fixed and frozen framework. This however is not quite as true of past centuries as we sometimes like to think. The nineteenth century did not invent the fact of change in the world. Great thinkers have always proclaimed the principle. Kant, Laplace and others have pointed out the law of continuous change in all things. It is a law which all must recognize. It is change from lower to higher, or from higher to lower, from simple to complex, or from complex to simple, in one endless interplay of things.

We know of course that the emphasis of yet more modern days has been upon the law of change. A passion for the idea of progress has possessed many minds, not simply a desire for progress, but an eagerness to detect progress everywhere. Both progress and fixity are true. No one can rightly doubt either. All practical science is based upon the certainty of change and the equal certainty of changelessness. The contents of the test tube are a wild whirl of forces. And they will be at exactly a certain level, to a hundredth of an inch, and with exactly such and such results, though all society be dissolved and kingdoms rise and fall. The star is an immeasurable gathering of powers so vast, of movements so incredible in speed, of temperatures so unbelievable, and of upheavals so appalling, that the mind cannot grasp them. And the star will be at a certain place at a certain day and hour and instant, to a hair's breadth.

61 Both change and fixity are true. How shall we relate change and changelessness? How shall we combine development with the conservation of energy? What is the universal reconciliation of these two? What is the universal law of progress combined with fixity?

Evolution Not this Universal Law of Progress

Herbert Spencer said that evolution was this universal law. He applied it to all things. But it is not a universal law in the sense which we are giving to that term. We need not discuss it as a process in species in animal and plant life upon the earth. That is a question of scientific data and of technical discussion. But Spencer lifted it out of science into universal application. He wanted to make it a law of universal progress.

But evolution, whatever its truth in species may be, is not a universal law, because it has no place in the universal things, the basic things, the fundamentals, of the universe. It is a conclusion from data which never can be complete or universal, which have to do only with species, and which do not apply to basic things. A universal law should lie deep in the being of space and matter and time, and evolutionary process has no place in these. No thinker would hold that space or motion or time or atoms or electrons were evolved by Darwinian or Spencerian formula.

Nor can evolution apply to the universe as a totality. It cannot be a law of the universe as a whole. For dissipation of motion is the basis of evolution. And dissipation of motion from the universe as a whole is incompatible with the fact of the conservation of energy in the universe as a whole.

Further, evolution does not mean universal progress. No one really holds it as such a law. Spencer did not. He said "Evolution" and "Dissolution." Those were his two factors. Progress and the opposite. Many scientists to-day say that "curve" is the law of evolution as they know it in their own fields. Things progress upward,--and downward again. Many scientists now teach "Devolution." That is progress downward. Men of science do not attempt to make evolution an invariable law of universal progress.

It is not an invariable law of universal progress in human life. For no one would say that progress is universal in human life. Individuals, as we all know, do not all make progress. Nations rise, but other nations fall. Civilizations grow, but civilizations decay. Arts grow, and decline. There is steady increase in knowledge, but this is because knowledge is accumulation, not growth. Many nations have progressed, but not all. Vanished Aztecs, perished Incas, the degenerate remnants of ancient African civilizations,--Asia, Africa and the Americas are strewn with immense evidences that progress is not universal in human life and society.

Evolution then does not mean universal progress. And further, it is not a universal law and cannot be a universal law, a law of the universal being of things, because it does not allow for the universal fact of changelessness. That fact of changelessness has been recognized from the beginning of thought as of equal weight with the fact of change. Evolution as a rule of change or progress means complete discarding of the old and the outworn, a vanishing of types left behind, an advance which obliterates all those factors which have been outgrown. It is not a universal law in a universe in which changelessness is as marked as change.

Above all, evolution is not a universal law, because it does not apply to God. He did not and does not evolve. Evolution cannot be a universal law, a law of the universe, when it does not apply at all to God. It cannot be universal to a theist. Spencer could not be a theist. It is not surprising that he felt himself obliged to eliminate God from the universe. It is not surprising on the other hand that others have made a grotesque attempt to apply evolution to God. Spencer's evolution, which could not apply to God, made the universe entirely different from its ground! But anything which is an absolutely universal law should be found also in God, who is the cause and ground of the universe.

Evolution, then, is not always progress, it is sometimes downward. It does not apply to the universe as a totality. Evolution is by no means universal in human society. Evolution does not allow for the great fact of changelessness. Evolution does not apply to God.

The Law of Changeless Change, or the Universal Law of Progress

It is true however that there is change and growth. That fact, deep in everyone's experience of things about us, is what led many to accept Spencer's attempt to lift a process of species to the level of a universal law of progress. Is it necessary that we should discover a universal law of progress in the physical and spiritual world? I do not think that it is. The secret of progress for the soul is God. We can know Him. But the question remains. Is there a universal process of change and progress? One which applies to all things? One which is self-evident in the basic things of the universe? One which agrees with the fact of fixity in the universe? One which includes God?

Yes. By going direct to the self-evident universal facts of the physical and spiritual world, we do find a universal law of progress, deep in the very being of all things.

It is the law of progress from source to embodiment, and from embodiment to contact or influence, from nature to person and person to personality, from energy to motion and from motion to phenomena, from future to present and present to past, from space to matter and matter to time. It is a law of continuous change and progress in the very nature of being and the very process of existence.

This is a universal law. It applies to all things. It includes all physical things, since they are of matter or motion and of time. It includes all personal or spiritual being.

This is a law which we can apply to the universe as a totality. For its continuous process does not at all require dissipation of energy from the universe.

This is a law applying to all human life, since it is the very, nature and process of human existence.

This is a law of the absolutely basic things of the universe. It is self-evident, not a matter of the gathering of partial data.

This is a law which includes God. It is the being of the Triune God. It is the reflection and the working, in His universe, of the Triune creative and ever-present God.

This is a universal and self-evident law of progress and change which does not do away with the fixity of the universe, nor affect the eternal changelessness of God. Eternal progress from Father to Son, and from Son to Spirit, does not mean leaving behind Fatherhood or Sonship in that progress. Perpetual progress from nature to its embodiment in the person, and from person to its influence in personality, does not mean a cessation of nature or of person. Continuous development from energy to its embodiment in motion, and from motion to its contact or influence in phenomena, does not mean an abandonment of energy or of motion. Constant procession of future into present and of present into past does not mean that in that process all future and present thereby immediately cease to be. It is a law of continual change with all the elements continually and endlessly existent. It is changeless change, in which endless development is a characteristic of endless fixity, and endless fixity contains within itself endless change.

This is the law of changeless change. It is change without dissipation of energy. It is a continuous change which discards no factors in its progress. It is self-evident. It does not depend upon partial data. It is the process of all the basic things of the universe. It is true of the whole universe of space, of matter and of time. It is true of man. It includes God.

The Law of Progress Practically Applie

This law of change and progress is one which can be applied consciously and practically to human conditions. It is a law which can be used to achieve definite ends. We cannot do so with the mere fact of change in the universe. It is not definite enough. We surely cannot so use evolution, since evolution is not always progress and does not always work upward, but often works downward, is not controllable, and does not always operate with men and nations. But we can apply the Divine process. We can apply the universal law of progress which in human life moves from nature to person, and from person to personality. We can apply it to human conditions and social questions.

This is Christianity's method of social progress.

First change the nature. Change the nature by regeneration.

Then by that regeneration change the person, his habits, speech, points of view, loves and hates, likes and dislikes.

Then from the change in the nature, through the change in the person, change the personality, the self as related to others, the contact, the influence, the environment.

This universal process of existence, this law of progress in the universe and in human existence, is the true method by which to achieve human progress. It is a universal, fundamental method. It is absolutely scientific. It is a method which can be used by anybody, a practical method. It is in harmony with God's way of working in the whole universe. It is accomplished by the simple fundamental method of letting the Triune God and His triune process into a person's life. It is very easily done, by old or young, by the intellectual man or woman or by the African savage. The person accepts the Son, the embodiment of the Trinity. He becomes a follower of Jesus. He opens the door to Him. The Triune God comes in. The nature becomes that of a child of the Father. As many as received Him, the Son, to them gave He the right to become children of God, who were born, not of flesh, nor of the will of man, but of God. The person becomes united with the Son, in a redeemed, risen, Christlike life. Then the Holy Spirit transforms, beautifies, glorifies the personality, and the influence upon others, and the whole environment. It is very simply done, as far as our part is concerned. It is the same in all lands and all cases.

By that method the new and wonderful Christian society arose in the midst of a frightful Greco-Roman civilization. The humanizing influence of Greece and the organizing influence of Rome had done their best. Art had surrounded the people. There ere no cities of such beauty to-day. A remarkable intellectual atmosphere was thrown around people. An extraordinary genius for government and order gathered the known world into the Roman organization. But society, amid all this humanizing and organizing environment, had grown steadily worse, and became one vast cesspool. The attempt to begin with environment had failed. Christianity came, and began in the Divine order, the method of God's universe. It began with men's nature, and changed that. As an inevitable result, by the universal law, the person became changed. His habits, his likes, his speech, grew different, and became like his changed nature. Then his changed personality was felt. It influenced his home, his friends, his community, the Roman Empire. This was the influence of Christianity, the influence of changed persons, with a changed nature, who had become children of God. It was and is a simple, universal method, allowing for a wide variety of personal experience. In some people acceptance of the Son is emotional. In some it is demonstrative. In some it is quiet. In some it is matter-of-fact. In some it is simply following that Person, Jesus, the Son. In some it is embracing Him under the passionate need of forgiveness. In some it is the throwing away of old doubts and the open confession of Him. In some it is the determination to live a life with His power in it. It is as broad as the human race. It is God's way and process. It is the work of the Triune God applied to social progress.

Laboratory Proofs of the Law of Progress

The entire failure of evolution as a law applying to human progress, and the fact that this Divine process is the effective one, are shown by laboratory experiments on a vast scale.

The first one is China. There are certain tests of a trustworthy experiment. One is universality. The experiment must include enough cases to make sure that it is not a sporadic instance. One test is repetition. The experiment must cover time enough to show again that it is not an accident. One test is isolation. Other factors must be shut away from the experiment. China meets all three of these tests. One-third of the entire human race. Four hundred million people now. There indeed is universality. An unbroken, full and careful record of her history, civilization and life for four thousand years. There is repetition. Completely isolated during these four thousand years, not by the Chinese wall, but by a higher, more permanent wall of custom, from contact and intercourse with other peoples and civilizations. China fulfills all the requirements of a vast and accurate experiment.

What is the outcome?

In four thousand years of unbroken history China evolved or progressed not at all, in government, in politics, in business, in science, in art, in literature, in morals, in religion. China has progressed more under the influence of the new, Divine method of Christianity in three generations, a hundred times more, than she had in four thousand years.

When you were a boy or girl in school you used to prove your examples in arithmetic by reversing the process. You can do that in this case.

The reverse proof is Germany. A very large, a very strong, a very self-contained nation. Sixty years ago Germany stood on the heights of civilization. A great, God-fearing, home-loving, law-obeying, intellectual nation. The nation of Luther, of Kant, of Bach, Beethoven and Wagner. But a generation ago a group of leaders, some of them scholars, some of them military men, some of them philosophers, set themselves to eliminate the second Person of the Trinity from Germany. They partially succeeded. They largely eliminated Him as Deity from the Bible, from many churches, from commerce, from education, especially from the universities, from politics, from the army. They eliminated the Divine factor and process. And in a generation Germany, with Him so eliminated, fell from the heights of civilization to that level in which she now stands. There indeed is the reverse test! Germany's is not the only case, but it is most striking. And Germany's unknown but true leaders to-day, in their desire to have their beloved Germany recover her soul, her conscience and her vision, and her place in the world, know that the only way to bring it to pass is God's process. German men and women must open their hearts to the lost Son of God. They must become children of the Father. They must get a new Spirit, and a new influence upon each other and upon the world. May the Triune God grant it!

That was Christianity's mighty and Divine method with slavery. It did not preach against slavery, in an attempt to begin with results. It changed the nature and the lives of slaves and masters, and made them brothers in Christ, and children of the Father together, and possessed by the same Spirit. And Christianity, by that Divine process, has destroyed slavery.

That was Christianity's method with womanhood. Woman was a slave, a plaything, without a soul, as she is in India and Africa to-day. Christianity quietly made her the equal in Christ, as a child of the Father, w as a temple of the Spirit, the full equal of man. And wherever Christianity went with this Divine method, womanhood has come to its own.

Civilization faces collapse, many say, perhaps rightly. The only way to stop collapse is the Divine law of progress.

If someone objects, "Your triune progress will not operate without God to work it," we answer, "We are not trying to find a way to make the universe or human nature operate without God. We are trying to find a process which works."

The true workman, the true thinker in action, is the one who will take the method of the universe and apply it to his own work,--and, if he needs to begin so, to his own life. He is one who will let the omnipotence of the Three in One come into his life and into his work. He is one who consciously lives in the triune kingdom of the Triune God.

VII. THE SECRET OF THE UNIVERSE AND THE PROBLEM OF ETHICS, OR THE GOOD

It should be no surprise to us to find that moral action is based in its structure on that Triunity which is the reflection of the Being of God.

Would that all moral action were based in its character upon the Being of God! But in its structure, whether its character is good or bad, all moral action, whether it will or not, is inevitably based in its procedure upon that universal triunity which is the structure and pattern of God's physical and spiritual universe.

For in the Motive, the Act, and the Outworking or Consequence, of every moral action or decision, will be seen in every way the triunity by which the universe reflects its God.

This is especially reasonable. For Ethics are based upon the Being of God. Ethical standards are not arbitrary. Moral Laws are not accidental. Goodness is not based upon some speculative, or over-abstract, or over-practical, or materialistic, or unreal principle. Goodness reflects the Goodness of the Creator. Holiness is what every personal being ought to have because God is holy. Conscience is the reflection in the human soul of the Holy Nature of God. Purity is the clear light of God in a human life, so that the "pure in heart can see God." God is more than holy. That He is "holy" is a characteristic. It describes God. But there are words which define God. They declare His absolute Being. "God is Light," absolute, stainless, shadowless, sinless, ineffable, glorious Light. "God is Love," absolute, unselfish, glowing, radiating, marvelous

Love. "Light" and "Love" are more than adjectives, describing His characteristics; they are nouns, defining His very Being.

And Ethics are based on that moral Being of God. That is why right is right, and good is good, and wrong is wrong, and evil is evil. God is the reason.

It is especially reasonable, then, that all moral action should in its procedure be based on the Being of God, which is so reflected in all universal things. Evil acts are of course a distorted reflection, a darkened, hateful, degenerate reflection. But in their structure all moral acts are a reflection. And good deeds are a glorious reflection. For good acts reflect not only His Triune Being, but His Being of Light, and of Love.

What is the structure of moral actions?

1st, is the Motive. That is the Source. Without the Motive there is no moral action.

The Motive is the Source. It lies deep in the soul. It wells up from unknown depths. It may even emerge from the subconscious. The object of the action may be from without, but the Motive which responds to it is from within. The Motive is unseen. It is invisible. It is the Source of every action.

2d, is the Act. The motive takes form in the Act. It is only as it does so, and becomes an act, that there is moral action. It may be inward, or intellectual, or emotional, or imaginative, but it must take some form, visible or invisible, to make it moral action.

The Act is the embodiment of the motive, in visible, or tangible, or audible form, or intellectual, or emotional, or imaginative form. Visible or invisible, the Act is the embodiment of the Motive.

The Act is the executive factor in moral action. It does whatever is done. The motive does things through the Act. Whether visible, or audible, or tangible, or invisible, intellectual, or imaginative, the Act embodies the motive, and does what is done, and makes moral action actual.

3rd, is the Consequence, or outworking, of the Act. It is the Act in its impact on other lives, or other things. Moral actions never take place in a vacuum. If they influence no one else, they influence one's self. No action can be judged, as to its moral character, entirely apart from its consequences.

It is not only the Act which works out in other lives, in other things, in environment, in one's self, through the Consequence. It is the Motive also working, through the Act, in the Consequence.

In all these aspects Moral Action is a vital part of that vast triunity by which the universe is in its pattern and structure the reflection of God.

All three of these factors in moral action must be judged together. The Act must be weighed by its Motive, and by its Consequences. So it is in any righteous court of judgment. The Motive must be judged by its outworking in Act and Consequences. The Consequences must be judged in the light of the Act itself, and back of that by its Motive. So it always is in honest, understanding comprehensive judgment, by every candid judge or observer. No one of the three,--motive, act or consequence, may stand alone. Each pervades the other two. Each permeates the entire moral action. Each has part in making the whole moral action good or bad.

Moral action is impossible without all three of these factors.

The causal order is always the same. The motive is always the source. The act is always the embodiment. The consequence always flows from the act, and from I the motive through the act.

It is a fixed, inevitable, invariable order, in absolute reflection of that Triunity which rules the universe.

Even the man who defies God, who hates God, who denies Him, who strikes at Him, does these things, in spite of himself, in a triune pattern and structure of activity which reflects Him whom he defies, hates, strikes at, or denies.

And acts of goodness, of love, of mercy, of unselfishness, of purity, of heroism, of patience, of humility, of sacrifice, how marvelously they reflect God, not only by their structure reflecting His Triune Being, but by their character reflecting Him who is Light and is Love. And the human life which is made up of such acts as these,--is there any other reflection of God so like Him, in all this universe as we know it?

So does the principle of Ethics become truly Divine. So it is lifted out of all trivial, or calculating, or self-seeking, or merely human, or mechanical, explanation, and becomes grounded on the Centre and Cause of the universe and of human existence.

And so does the Good become profoundly, not only in its moral character, but in its very pattern of action, the reflection of God.

VIII. THE SECRET OF THE UNIVERSE AND THE PROBLEM OF REALITY, OR THE TRUE

[**]How the mind sees things--Why things are universals and particulars and things-as-related-to-others--Which one is real?--The forms of pure reason. Why they are what they are--Forms of thought versus outer realities--Which are more real? The mighty answer.

This Triunity underlies the forms of reality.

Human reason sees things as percepts,--that means as existences perceived, as things as they simply are, as things in themselves. It sees them also in a second way, as concepts, as general types.

Then still further it sees the things as related to other things.

It sees, for instance, the individual man. It sees, in addition, humanity in general in him. It sees, finally, the man as related to others.

Reason sees, for further instance, the particular tree, simply as a tree, the thing in itself. It sees also the universal idea or type or fact of tree nature. It sees also the tree in its environment, as related to other trees, to other things, and to all the universe.

We have clearly three things here. First, the percept, the particular thing which we see. Second, the general nature or type, of which the particular thing seen by us is an embodiment. Third, that particular thing as it is related to other things.

Why do things exist in these three ways? Or, if the reason for it is not in the thing seen, but in the mind, why are our minds such that we must see all things in exactly these three ways? Whether they are the forms of reality as they exist in themselves apart from us, or o only the forms of reality as we see them, why are they as they are? What is the underlying basis of them?

And which is the more real,--the thing in itself?--or the universal type of which that thing is a particular embodiment?--or the thing as related to other things which it touches and to all other things?

This is the third great problem of thought through the centuries, beginning with the most ancient days of Greek philosophy. It is very plain that these three things,--the thing in itself, the general type and the thing as related to others,--coincide with the three distinctions in the universal triunity in human existence, or in matter, or in time.

The person is the particular thing which I see and know. Then on the one hand there is the nature of which the person is the embodiment, an individual nature, but shared with all other human beings as a universal human nature. And on the other hand there is the personality, which is the person as related to others.

Or we may, before we go further, put it in the triune order of human existence: first, the universal human nature, of which the individual person is the embodiment; second, the particular or individual person; third, the personality, the person as related to others. That is:-- the universal type or nature, the particular embodiment or thing, the thing as related to others. So also of matter. Energy is the universal, the source, the unseen. The particular motion is the embodiment of that energy. Phenomena are that particular motion in contact with other existences.

So it is also of time. The future is the universal source, the potentiality. The present is the particular realization, the embodiment, of the future. It is the thing we know and touch. The past is that present as soon as it has related itself to other things.

All of this is very clear. The progress from nature to person, from energy to motion, from future to present, from source to embodiment, is a progress from the universal to the particular. The progress from person to personality, from motion to phenomena, from present to past, is a progress from the thing in itself to the thing as related to others.

We have then this triunity: 1. The universal. 2. The particular thing. 3. The thing as related to others.

Why Things Are Universals and Particulars and Things-as-Related-to-Others

What is the basis of these realities? Why do things exist as universals and particulars and things as related to others?

Because they are nature, person and personality. Because they are energy, motion and phenomena. Because they are future, present and past. Because they are triunity.

Because they exist in the image of the Triune Reality and Ground of the universe.

Which One is Real?

Which then is reality? the universal? or the particular thing? or the thing as related to others?

That is indeed a much debated question. And the conclusion of the debate depends upon the angle from which one looks upon the question. One comes out where one went in. One may be an idealist, and emphasize the universal. One may be a realist, and emphasize the particular thing. One may be a pragmatist, and emphasize the thing as related to others.

But in the light of the triune reality of the three, the answer is clear at once.

"Which is more real? the universal? or the particular thing? or the thing as related to others?"

Each is real, in that triune reality, and each is dependent upon the other two for its reality.

The nature is real, but not apart from its embodiment.

The embodiment is real, but it cannot exist without its nature which it embodies.

The embodiment, the thing in itself, is real, but it cannot be real without coming into contact with other things and becoming the thing as related to others.

If the embodiment cannot be real without the thing as related to others, clearly the nature, which cannot be real without its embodiment, cannot be real without the thing as related to others.

And manifestly the thing as related to others cannot exist unless it is first the thing in itself, and, back of that, the nature. Each is real. Each requires the other two for its reality. Each is the whole. All are real, in the light of the triune reality, in the image of the supreme Triune Reality and Ground of the universe.

The Forms of Pure Reason. Why They Are What They Are

The strict forms of pure reason, the formal expressions of reality, which we call deductive logic, are equally an expression of the same underlying triunity. We have only to examine the syllogism to see that in all its forms it is what it is because of that universal pattern.

The major premise is the universal. It is the source. Out of it the syllogism grows. It is the fundamental truth. If it is not fundamental it cannot be a major premise. It is the nature of the thing under discussion.

The minor premise is the embodiment in a particular form of the general nature, the universal principle, in the major premise. It sets forth that universal principle in a specific, particular form.

The conclusion proceeds from the minor premise. It proceeds from the major premise through the minor premise. It brings both major premise and minor premise, both the universal principle and its particular embodiment, into contact with the things under discussion. It applies the syllogism to life and conduct and environment.

The nature or universal, the major premise,--the embodiment or particular, the minor premise, the conclusion, or application of the major and minor premises, of the nature and the embodiment, to the things of life and conduct,--these are the three invariable factors. The syllogism, the forms of pure reason, are identical with triunity. They are what they are because triunity in the likeness of the Three in One is the structure of the universe.

Forms of Thought versus Outer Realities

We are ready then to answer in the light of Triunity the essential question about these things.

Are the universal and the particular and the thing-, as-related-to-others forms of outward reality which exist apart from our thinking? Or are they simply mental forms under which our minds conceive reality? Are they forms of reality, or forms of thought? Thinking, ancient and modern, sways to and fro upon that question.

The same question comes in regard to space and time. Are space and time simply mental forms under which we conceive reality? Or are they themselves realities of the outer world?

Ancient philosophy held that space and time are outward realities. Modern philosophy, following Kant, tends to see space and time as forms of thought, through which we conceive the outer world. The modern philosophy which desires to transform itself into psychology is very sure of this. It sees space and time as purely products of the mind.

They are surely forms of thought. We cannot think of the outer world at all without conceiving things in terms of extension, or space, and of consecutiveness, or time.

We can think only in universals, and particulars or things-in-themselves, and things as related to others.

We can reason only in major premises, and minor premises, and conclusions, in some of their myriad forms.

They are necessary forms of thought.

And they are outward realities. For motion, which we all acknowledge to be a universal outward reality if there is any outward reality at all, apparently cannot take place except in actual space. But if that be doubted, we see motion now definitely, in this new world of triunity, as the result of space. We see it as the motion of space, the outspreading of power, emerging through energy into motion. Motion is not a reality unless space, the outspreading of power, is real. And on the other hand motion cannot take place without generating time.

And as for universals and particulars and things as related to others, we have seen that they are bound up in one reality. We cannot take out one of the three, and say "This one, and this one alone, is real." In their triune relationship and existence, all are real.

Space and time, then, and universals, and particulars, and things as related to others, are both mental forms and outer realities. They all exist in the image of the Triune God who is the Reality and Ground of the inner universe and of the outer universe.

Which Are More Real? The Mighty Answer

Which is more real, the inner world with its forms of thought, or the outer world with its motion and substance? That question, great as it is, loses its meaning in the light of the Triune Reality. For both the inner world of thought and the outer world of motion and substance are based on the supreme Triune Reality.

In which then are space and time more truly realities? To which do space and time more truly belong? The question disappears in the light of the Fact of triunity. When we see space and time in their most fundamental aspect, as reflections of the Creator, and see the soul also, and all its forms of thought, as the exact reflection of the same Creator, it makes little difference to which space and time most belong. In both the outer world and the inner soul they are the one image of the Triune Creator.

What is true of space and time is true of universals and particulars and things-as-related-to-others. How largely do these forms of reality exist apart from us, and how largely are they the forms in which the mind works in seeing reality? The answer is truly clear. We can be sure that they are with equal certainty forms of outward reality and forms of thought. For both the world of outward reality and the mind with its forms of thought are made in the reflection of the Triune God.

What is the relation between the mental conception and the outer reality?

Does the outer world suggest Space and Time to the mind?

Or does the mind project Space and Time upon the outer world?

There is a greater answer. The Triune Creator suggests and projects Space and Time upon both the outer world and the mind, and together the outer reality and the inner conception form one operation of the Triune

God who forever creates both world and mind in His own triune likeness.

In this is the unity of the mind with the outer world. For He made both in His own likeness.

In this is the reason that the forms of reality and the forms of thought exactly fit each other. For He made both in His own likeness. This is why the mind can know the universe around it. For He made both in His own likeness.

For both together in all these things are one great image of the Triune God.

^194:* The general reader may, if he so prefers, omit this chapter, which is unavoidably technical in much of its language, and may continue the reading with the beginning of Chapter 9, on page <page 203>.

IX. THE SECRET OF THE UNIVERSE AND THE PROBLEM OF AESTHETICS, OR THE BEAUTIFUL

The unvarying factors in human creative work--"Why are they?"--"Which is most vital?"--The problem of Aesthetics. "Where is beauty?"--"All are right"--The Formula of the Universe.

It is reasonable to ask, "Does the being of the Creator, which explains so much about His creations, and especially about that wonderful creation, man, explain the creative work of man himself?"

Certain factors are always found in all creative work of man. Always there is and has been debate about them. Why are they as they are? What is the principle of them all? What is the unity of them all?

And there is the further question. Which of these factors is the most vital? It is an endless question, pro and con. The factors themselves are known to everybody. They are found in some degree in every creative work.

1st. The source,--the idea, the conception,--the ideal,--the inspiration.

2nd. The embodiment, the picture,--the poem,--the sonata,--the song, the statue,--the building.

3rd. The picture, poem, song, as it affects and moves others.

It is always these three factors, whatever the creative work,--whether Hamlet or a children's tale, whether the Last Supper or a sporting print, whether Tristan and Isolde or a folk-song, whether the seated figures of the Parthenon or a popular statuette, whether the Taj Mahal or a vine-clad hut,--always these three!

Why are they as they are,--always these three,--the source or idea, the thing which embodies the idea, the thing working in the souls of others?

Because the universe is so. Because matter, and time, and man, and space-matter-time, and the process of existence, and the principle of progress, and the principle of reality, are so. Because the Creator and Ground of the universe, of space, matter, time and man, the Creator of the creative energies of man,--is God the Father, the Almighty Source,--and God the Son, the marvelous Embodiment,--and God the Spirit, who moves in the souls of men.

Which of these three factors in man's creative work is the most vital?

That is the basis of vast discussion. Whole schools of art have arisen from this or that emphasis upon one factor or the other. Whole schools of theory have hung on this factor or that. Which is the most vital factor?

Some say the idea. And truly we must have the idea. Without it there is but an exhibition of technique. We all know the things without an idea. The galleries of dead but unburied paintings. The machine-made popular songs. The verse which a generation ago was sounding brass and in this generation is tinkling cymbal. The endless streets of dull or smart complacency. We must have the idea, the inspiration.

But we must have power to embody the idea. Else it is not art. It is impulsive amateurism. Many fail in that way. We must have technique to embody the inspiration, in well-painted picture, in well-wrought poem, in well-woven sonata, in well-modelled statue, in well-proportioned building.

But the picture, the poem, the song, the statue, the building, must touch and influence others. What matter how beautiful it is, if it has no spell for the souls which see or hear it! What use its power, its idea, its technique, if it leaves all other minds cold! We must have the idea,--we must have the technique,--we must have the instinct for other minds. The picture, the poem, the song, must live in other lives.

Which of these is most necessary? Which is most vital?

The answer lies deep in the nature of things. In the being of God,--the ground of the universe,-- Father, Son and Holy Spirit are so deeply One that no one of the three can exist without the other two, and no two can exist without the third. The three dimensions of space are so much one that no one of the three can exist without the other two, and no two can exist without the third. Energy, motion and phenomena in matter are so much one that no one of the three can exist without the other two, and no two can exist without the third. Future, present and past, in time, are so much one that no one of the three can exist without the other two, and no two without the third. So also nature, person and personality in human existence are so deeply one that no one of the three can exist without the other two, and no two without the third. And so also it is of human creative work, that no one of the three factors in it can exist without the other two, and no two without the third. It is the unity of life, and all three are vital, for the three are one life. With any of the three lacking, there is no life and no art. The uninspired creation is dead from its birth. The inspiration poorly wrought out is an ambitious failure.

The work of art which makes no contact with other minds might better have never been born. But when the three cooperate, there is life and victory. The inspired idea, the source,--working through the perfect embodiment, the visible, audible reality,--enters into its mighty influence, its living presence, in the souls of others,--and the Madonna di San Sisto, the Paradiso, the Missa Solemnis, the Winged Victory, or the Cathedral of Amiens, has fully come into the world.

The Problem of Aesthetics. Where is Beauty?

And this principle rises also into the realm of universal reality. It leads to the problem of Aesthetics, a universal problem, above and beyond all individual works of human creative art.

Where is beauty? In what does it reside? That is the problem of Aesthetics.

Some say that beauty lies in the ideal, the abstract, which the visible or audible object embodies. That is the view of the Platonist. He sees the ideal as existing in all its perfection, bright, ineffable, never wholly to be touched, above and before its appearance in any individual embodiment of it. This is what the Idealist holds.

Some say that beauty lies wholly or mainly in the object, the work of art or of nature, the statue, the symphony, the tree, the sunset, which we see or hear. That is what the Realist says. Beauty seems to him wholly objective. The ideal is to him something which we construct from the definite things of beauty which we see or hear. He is not sure whether the ideal truly exists. Certainly the specific thing of beauty is the most real to him.

Some say that beauty is in the mind of the beholder or hearer. Certain things give him pleasure. Certain things give him delight. These he calls beautiful. Beauty then, he is ready to say, and to demand that we should admit, is in the mind of the beholder. It is a purely subjective quality. That is what the Romanticist may say. It is what the Pragmatist does say.

Which is right? Where does beauty reside? In the ideal? Or in the embodiment, the object? Or in the mind of the beholder?

Which is right? And what is the reason for it? All are right.

Beauty resides in the ideal, the abstract. We can easily test it. We cannot follow the processes of the Divine Creator of the sunset, the mountain height or the flower. But we can follow our own processes. The artist or the artificer who would create an object of beauty, a picture, a song, a sonnet, a vase, without an ideal glowing before him, a vision of what he would like to embody in his work of art, will fail to capture beauty and fix it there. Beauty dwells in the ideal.

And beauty dwells in the object which we see or hear. If it is not there, the beholder or hearer will surely never know the ideal which is dimmed and concealed by the unbeautiful work of art. The artist must have his vision, it is true. But he cannot show it to us except in the beauty of the work which embodies it. Beauty dwells in the visible object, the work of art or of nature.

And beauty lives in the mind of the beholder or hearer, and would have neither meaning nor reality without its place in the mind of the beholder or hearer. There must be the thrill. There must be the delight.

There must be the pleasure. There must be the emotion. If there is nothing of these in any mind as it sees or hears the work of art, where for us is the beauty? How has it any reality? Beauty dwells in the soul of the beholder or hearer.

Beauty dwells in all three,--in the ideal, in the individual embodiment, in the mind of the beholder. And no one of the three can be the home of beauty or sublimity without the other two.

Why does beauty dwell so in all three: the ideal, the embodiment, the mind of the beholder?

Because this universe of beauty takes its character from the Creator and Ground of the universe, and reflects the beauty and sublimity of His being. He is the Ideal, the Father, the Source, revealed by the visible or audible embodiment. And He is the visible one, the Son, embodying the Father. And He is the Spirit, who moves in the hearts and minds of others. And all Three are One, in an infinite, intensive, almighty Unity.

Goodness, Truth and Beauty

These three ideals or facts mean much to many thinkers today. Here we see them in the forms of Ethics, Reality and Esthetics. Each of the three, Goodness, Truth and Beauty, is a perfect reflection of the Divine Triunity of Father, Son and Holy Spirit.

The Formula of the Universe

The universe of matter, and of time, and of man, and of man's creative work, and of sublimity and beauty, and of space-matter-and-time, and of the processes of existence, and of change and changelessness, and of reality, is one universe, truly a universe, with one pattern, one organic law, built in the likeness of its Creator,--Father, Son and Holy Spirit,--the Three in One.

And if to say that the universe reflects its Creator seems to any highly sophisticated mind too simple or too romantic, and something more abstract seems desirable, put it in this way!--that this which we have been discovering in one realm after another is clearly the formula of the universe;--and that this formula naturally and inevitably coincides with the principle of the being of God, who is the Ground of the Universe.

CONCLUSION. THE SECRET OF THE UNIVERSE AND THE RIDDLES OF THE UNIVERSE

The explanation of the things of the universe--Which explains which?--The reason of the universe--The Divine Method of Work--The Deeper Mysteries--The Riddle of the Universe and the Secret of the Universe.

The Trinity, imaged in the universal law of Triunity, explains the deep things of the universe. It shows why space is what it is. It shows why matter is what it is. It shows why time is what it is. It explains why human existence is exactly what it is. The Trinity, imaged in the universal triunity, is the basis of unity in all things. It shows that unity lies, not in a common substance, which is impossible, but in a common structure and pattern. It underlies the relations of space, matter and time. It shows space as potential motion, motion as embodied space, time as successive motion. It shows what is the vast and true relativity among them. It shows the infinite circuit of the universe, out from the mind and power of God, through space, motion and time, back into the mind and eternity of God. It shows the process of existence, the same in all things, and shows that there is no conflict between "being" and "becoming," because "being" is, within itself, "becoming." It shows the law of progress and of change and fixity in the universe, and the method of human progress. It shows the procedure and pattern of moral action, and the basis of the good. It shows the forms of reality, or the true, and why they are what they are, and why the process of reason is what it is. It shows the nature and reason of the beautiful. Wherever there is a universal thing, there, apparently, is triunity, and always with the same relations and characteristics. Triunity in the likeness of the Three in One is the structure, the pattern, the unity, the process, the progress, the reality, of the entire universe. The Triune being of God is the mighty solution of the riddles of the Universe.

This is only as it should be. The being of God ought to be the basis of the basic things of the universe. It ought to explain the problems of the universe. It ought to make clear many things which the mind by itself cannot settle. But until that being of God is seen in its full nature, as Jesus and the Bible reveal it, as Triune, the vision of it does not explain, it has never explained, these universal problems and mysteries. What we have done is an obvious thing. We have realized that when the being of God is recognized in its marvelous Triunity it explains all of these universal things at once and self-evidently. In so recognizing that we have here the solution of supreme problems we need not claim supreme minds. Anyone may gaze on the stars or see the sunrise. The true attitude is humility. What we need is not supreme minds, but a supreme Fact. There is no intellectual conceit in realizing that here is the key to the riddles of the universe. Why should not the being of God be such a key? It is as it should be. The central fact of the universe, the being of the Creator, the Three in One, explains the great things and illumines the otherwise difficult mysteries of the universe.

Which Explains Which?

The Divine Triunity explains these universal things. But these things do not explain that Triunity. The structure of the universe, the nature of space, of matter, of time, of human life, attest the Trinity. They reflect the Trinity. They demand the Trinity. But while they do all these things, they do not explain the Trinity. The Trinity explains them.

Someone will try to turn this whole revelation around, and argue that Father, Son and Holy Spirit are the effort of some long-ago thinker to put this universal triunity into theistic form. It will be so argued by some one who does not personally know the Triunity of God.

The answer to such an effort is, as we have said before, overwhelming.

1. There is no sign of such an origin in the Biblical presentation of the Trinity.

2. No man or men could build out of this universal triunity such a matter-of-fact, and natural, and almost casual, and wholly untheoretical presentation of Father, Son and Holy Spirit.

3. There is no reason to think that any man in New Testament days knew or could know this scientific, universal triunity.

4. Above all, such an explanation does not explain this universal triunity itself, in space, in matter, in time, in all three together, in human existence, in human self-realization, in human self-direction, in the laws of reality, in the process of the universe. This universal triunity has a cause, if this is an orderly universe. If this is a theistic universe, that cause must be in God. Such triunity in God is not the result of the universal triunity. It is the cause.

For that is what the universe demands as its explanation.

It demands such a Triunity in God, as the cause of the universal triunity. And that is what the New Testament Triunity demands as its explanation. It can have come only from the Triunity which the whole universe reveals in God.

The universal triunities, therefore, in space, matter and time, and in other universal things, do not cause and do not explain the Divine Triunity.

But the Divine Triunity alone could cause and alone explains these universal triunities.

The Reason of the Universe

Is the universe as it is because of some special plan which requires that the universe should be so?

Or is the universe as it is from some inherent necessity?

These great questions disappear in the light of an answer deeper and greater than either. It is not simply a special arbitrary plan, chosen out of endless possibilities. Nor is it on the other hand simply a necessity in the nature of the universe itself. It lies far deeper. The universe is as it is because it naturally and inevitably reflects the being of its Maker and Worker. That means indeed a plan of the universe, but not an arbitrary plan. It means a necessity, but a necessity far deeper than anything in the nature of the universe itself. The universe inevitably reflects the being of its Maker and Worker. He creates it upon lines of His own being. He perpetually creates it and works in it in harmony with His own being. He expresses His own being in it. He shines through its processes. It is the visible vesture conformed to His own mighty being.

The Divine Method of Work

All these triunities are the workings of an immanent God. He is not merely the Creator. He works now in His universe. These things are His method of working. They are not static. They are not fixed and hard. They are His living method of working.

The relation between energy, motion and phenomena, or between future, present and past, or between space, matter and time, or between nature, person and personality, is an active relation. It is not merely an architectural relation between the three in each of these triunities. It is an active relation. These things are not buildings. They are processes. It is a working relation. It is the immanent God, working through these methods. He works from energy through motion and phenomena. He works from future through present and past. He works from space through motion or matter and time. He works from nature through person and personality. They are His constant and active method. He works through these triunities in His universe. They are not merely a passive reflection of Him in a fixed and universal mirror. Your mirror reflects you. But far better and more truly your ways of work reflect you. These triunities reflect God not only as the passive mirror of creation reflecting the Creator. They reflect Him as your ways of work reflect you. They are the triune methods of the Triune God working in His universe. As Creator He is reflected in them. As the Worker in His universe He shines through them.

The Deeper Mysteries and the Universe

So all-explaining is the light of the Divine Triunity of God in His universe, that even the deeper mysteries of that Triunity cast a revealing light upon the mysteries of the universe.

What holds the universe together, so that it works as one immeasurable whole?

What holds the stars in their order and harmony? What keeps them in their orbits

What holds the atoms in order? What holds the electrons in their orbits around the proton in the infinitesimal solar system which we call the atom?

The only answer which has ever been given at all is the answer of the Bible, that "in him,"--the Second Person of the Trinity, the Son, the Creator,--"all things hold together," or "consist."

What holds the mind together, in the yet more wonderful inner universe? What holds intelligence, and feeling, and willing, and memory, and imagination, together in order and harmony in the mind? The only answer which has ever been given, or ever attempted, is the answer of the New Testament, that "in him," the Son, "all things hold together."

It means that the Creator, the Son, holds atoms, stars, forces of nature, forces of the mind, "things visible and things invisible," the whole vast universe, together in order and harmony, in life and unity.

If this mighty answer is true,--and certain it is that no other answer has ever been given, then the universe centres in the Son. The same New Testament which brings to us that Divine Triunity of Father, Son and Holy Spirit which the universe requires depicts also the universe as centering in the Second Person of that Triunity.

But if the universe centres in the Second Person of the Trinity, should not the reflection of the Divine Triunity in the universe be primarily a reflection of the Son? Would not that be strange?

Yes, but so it is. The image in each of those great triunities which make up the universe is above all an image of the Son! The emphasis is upon Him in all of these universal reflections of triunity. Nature and Personality centre in the Person. They are both invisible. It is the Person which we see and know. Future and Past centre in the Present. It is the Present which alone we can touch and know. Energy and Phenomena centre in Motion. Space, Matter or Motion and Time centre in Matter or Motion. Space and Time we know only through Motion or Matter. The second factor is the most vivid in each triunity. The second factor is not greater than the other two, but it is the most vivid, and so the first and third elements centre in it. It is motion, in energy, motion and phenomena,--it is the present, in future, present and past,--it is matter or motion, in space, motion and time, it is person, in nature, person and personality,--which is central and most vivid. And now we see the reason. It is not because the Son is greater than the Father or the Spirit. It is because the universe in its vast triunity reflects most vividly the Second Person in the Three in One, the Son, the Creator, "in whom all things consist." A yet deeper mystery the Bible declares in regard to the Divine Triunity. And yet it fits the facts of the universe and casts light upon these mysteries of the universe which we know only too well. The Bible declares, in a mysterious passage, that "at the end" "the Son shall deliver up the kingdom to God, even the Father." This is to be when He, the Son, "shall have abolished all rule and all authority and power." It alludes to the power of every evil authority or force, and to the power of death. "For he," the Son, "must reign till he hath put all his enemies under his feet. The last enemy that shall be abolished is death." This evidently means all that force of death and destruction, both in human life and in the universe at large, which negatives God's whole creative purpose and work. "And when all things have been subjected unto him," the Son, "then shall the Son also himself be subjected to him that did subject all things unto him, that God may be all in all." A profound mystery, but a profound illumination! We see it going on now. Sin, disorder and destruction permeate the universe now. The Son, the Creator, Himself enters the life of the universe in a peculiar and personal way. He does it by entering, as a Person, the life of the human race. He becomes man. He overcomes sin. He does this in His own life for thirty-three years. He does it for mankind, the Bible declares, by His Personal death and resurrection. Then He works it out, in men and in the universe. He works in men, the key to the whole vast situation, by His Spirit. At last, "at the end," He brings all things into subjection. A new-created race emerges, from every tribe and tongue and people and nation. A new universe, no longer groaning and travailing in pain. New heavens and a new earth. "All things new." Sin and evil and destruction cast out from it all. Then at last all things are in harmony with God. The Nature of God at last holds absolute sway in the universe. The redeeming, reorganizing, recreating work of the Son is done. No longer must one of the Three in One make it His work to reclaim the universe from sin and disaster. God, the Three in One, the One in Three, is at last all in all in His universe.

But now, as we can see everywhere in the universe, as we have been seeing in one realm of the universe after another, it is the Son in the Divine Triunity who is above all reflected. It is He who is reelected in the vast interwoven fabric of Motion. It is He who is mirrored in the living Present. It is He who is imaged in Persons. And it is through Jesus, the Son, that we came to know that whole Divine Triunity of which He is the Second Person. He was our point of contact with the Three in One. It was His claims, backed up by His character and personality, that brought the Divine Triunity to us. To know Jesus is to know the Triune God. Anyone who knows the Triune God will agree to that. And to know Jesus is to know the Secret of the triune universe. He is the key to the great mysteries and realities of God, and the great mysteries of the universe, of space, of matter, of time, of the relations of space, matter and time, of human existence, of the process of all existence, of the law of all progress, of the forms of reason, of "being" and "becoming," of the unity of all things, and of countless mysteries yet to be revealed. He is the key, "in whom are all the treasures of wisdom and knowledge hidden."

The Riddle of the Universe and Its Almighty Answer

The Riddle of the Universe brings its own universal Answer. The Riddle leads to Reality.

"Why are all things what they are?" becomes "Why are space, and matter, and time, and space-matter-and-time, and human existence, and progress, and moral action, and reality, and beauty, all triune, in exactly the same sort of way?"

And by its universal corroboration and its universal demand the Riddle leads directly to its own Answer in the Triune God. That is as it should be. What else but the being of God could explain His universe? He is the Cause of all its triune structure. He is the Worker in all its triune methods. He is the Solution of all its triune mysteries.

Who knows what wonders we may yet discover, beyond all the wonders of modern science, in the natural world and the inner world, when we have learned to see and interpret the universe in the light of its Triune God?

How many decisive formulae, to unlock new resources of power, lie undiscovered in the triune Formula of the universe?

How many far-reaching principles may radiate from the triune Principle of the universe?

How many processes, of value to men, may reasonably be produced from the Process of the universe?

And surely the more we penetrate into the secrets of personal being, in our intense modern study of human life, the more we must see as their Secret the Triune Being who upholds the universe and in whose likeness and reflection man is what he is.

We may escape the danger, which now threatens us, that our immeasurably growing knowledge of the physical universe may overwhelm us,--if only we will learn to see the natural world in the light of its Triune God.

We may escape, too, the greater danger of the present day, in our over-eager study of our own being, our actions and reactions, our behaviour, our thinking, our reason, and everything else about us, that we shall analyze ourselves into conceit, inbreeding and ineffectiveness, and the deification of man,--if only we will see human existence always in the light of its Triune God.

Is it too much to say, that all things lie open to the thinker who knows the Triune God, and who dares to apply the supreme Fact of the universe to the other facts of the universe? And what greater things are open to him who applies it to his own life? There lies indeed the way of vision and power, for life is greater than thought, and to know truly the Triune God is life indeed.

The End

18394396R00072

Printed in Great Britain
by Amazon